Y0-BSD-037

Maximizing Forest Product Resources for the 21st Century

▲

NEW PROCESSES, PRODUCTS, AND STRATEGIES FOR A CHANGING WORLD

Richard F. Baldwin

SAN FRANCISCO

Published by Miller Freeman Books
600 Harrison Street, San Francisco, CA 94107

Cover Design: Brad Greene
Interior Design: Brad Greene

Photo credits: page 9—Erik Beard; page 15—Dave James; pages 21, 85 and 96—Raute Wood; page 56—Bill Carter; pages 148 and 149—Temperate Forest Foundation

Library of Congress Cataloging-in-Publication Data

Baldwin, Richard F.

Maximizing Forest Product Resources for the 21st century: New processes, products, and strategies for a changing world/by Richard F. Baldwin

p. cm.

Includes bibliographical references (p. 195) and index

ISBN 0-87930-599-1 (alk. paper)

1. Forest products industry—Forecasting. 2. Lumber trade—Forecasting. 3. Forest products industry—Environmental aspects. 4. Lumber trade—Environmental aspects. 5. Forest products industry—Technological innovations. 6. Lumber trade—Technological innovations. 7. Forest conservation. 8. Forest products industry—United States—Forecasting. 9. Forest products industry—Canada—Forecasting. 10. Lumber trade—United States—Forecasting. 11. Lumber trade—Canada—Forecasting. I. Title.
HD9750.5.B35 1999
338.1'7498—dc21 99-43322
CIP

Printed in the United States of America

00 01 02 03 04 05 06 5 4 3 2 1

Contents

Preface

■ The forest industry is an essential global industry—yet it is an enigma. Some believe the Paul Bunyan heritage still lives: a heritage of big men, huge mills, and ancient trees. Others are not so sure; they know trees are the mainstay raw material for lumber, paper, and other products, but they don't know if the forest reemerges after harvest—and if it does, will the land ever return to its preharvest condition. Somehow the industry has not escaped its cut-and-move-on reputation of the 19th century.

The consumer and the industry have somehow each failed to recognize change in the other. Past consumers recognized the land and the forest as a producer of crops; today's consumer expects far more—wild and scenic vistas with abundant wildlife hold major importance, and high-tech industry is the favored job creator in preference to basic smokestack businesses. Yet the industry is changing in ways unimagined even a few decades ago.

High tech has arrived; it has arrived in how the industry grows its trees, manufactures its products, and how these products are used. Billowing smokestacks have been replaced by well-regulated steam and air emissions. Teepee burners, those that are left, are cold, colorful, rusting relics of the past. The wood waste that fed those burners is now the favored raw material for a variety of products.

It is not uncommon to have high-tech forest industry manufacturing facilities located within landscaped grounds adjacent to sophisticated facilities making products unimagined a generation ago. And we really haven't seen anything yet as the forest industry emerges into the new century. Certainly, there will be remnants of the past still operating, in both North America and other parts of the world. But these remnants will be fewer and further between as the new century unfolds.

And how about the forest? Again, the observer hasn't seen anything yet as the raw material demand is fed by fiber farms, urban waste, and other unlikely sources. Will trees still be grown and harvested in the forest? A resounding yes! But the grower and harvester will be sensitive to the consumer, a consumer who demands something more than logs from the forest, even on private lands.

Three generations of Baldwins have witnessed and participated in many of the major industry changes since the late thirties. The third and later generations will witness many more. This book, then, is a progress report to the industry participant, the consumer, the investor, and others who depend in some way on the forest and its products. It briefly describes the past, and then moves on into the future, a place we will spend the rest of our lives.

The book is organized into five sections; the first section, An Overview, is a three-chapter installment that establishes the foundation for the sections that follow. This initial segment describes the significant events within the industry as the 20th century wound to a close, the springboard for the changes that follow. Dynamic raw material and environmental issues are described. A chapter on how research and development will provide solutions describes the emerging response to growing consumer expectations, a shrinking working forestland base, and a host of other issues.

The second section, The Raw Material Base: Globalization and Re-engineering, describes evolving and changing manufacturing technologies and the wood products that develop from these technologies. These chapters highlight the specifics of what is happening in the forest, in the laboratory, and in the gathering and reuse of wood fiber to meet raw material needs.

The Mills: Adapting to Change is the title for the third section. An in-depth description of how newer wood-based products are manufactured to materially improve upon nature provide the producer with knowledge on how they can improve the company prospects for growth and profitability. But it doesn't stop there. The importance and application of adhesives systems is detailed, along with how-to advice on overcoming the complexity in dealing with people, products, and processes. The section then cites specific ways to become a low-cost producer.

Marketing and sales strategies are covered within Section Four. This section describes the emerging population and consumption trends, then follows with strategies and tactics to meet these trends. Forest certification is described at length. Bottom line, the consumer is having an increasing say in how the raw material is obtained, how the product is made, and how the resulting product is accepted and used.

The final section, Moving Ahead to the 21st Century, focuses on leadership; specifically it addresses the issues of credibility, responsibility, and effectiveness in dealing with people. These chapters outline the lessons of the past and provide specific strategies and tactics to be used to benefit from the compelling and unprecedented changes that are rapidly unfolding

Does this book have a solution for every question? Probably not. I wish it did. But it does benefit from the author's over 40 years as a participant and observer, the last 30 as a published author. Hands-on business activities, interviews, observations, and detailed research in the United States and 20 other countries covering five continents over the decades have provided material for the text. Specific examples come from both the author and others. And who are the others?

That's a tough one, because there are so many. The author has credited specific individuals within the text, a comprehensive bibliography is also included that identifies the host of helpers and contributors. Whenever possible, the author not only reviewed the cited references, but followed up with engaging and compelling interviews. But there are individuals who are credited neither in the text nor in the bibliography. Miller Freeman's Kathy Porter is one.

Kathy and I met at her Atlanta office to discuss the preliminary outline for the book. Her suggestions, her enthusiasm, and her encouragement played a large role in the way the book was written. Also, thanks to Miller Freeman's Dorothy Cox and Mike Kobrin, who handled the gargantuan task of editing and keeping me to deadlines.

Then there is Betty Baldwin, the mother of our five daughters and three sons. Since 1969, Betty has continued to be a major contributor to the many books and articles written by the author. She did a yeoman's task proofreading, critiquing, editing, and working with the Miller Freeman production group on this most recent book.

Sempelkamp's Fred Kurpiel is another. Fred is the Vice President of Sales in the company's North American headquarters in Atlanta. He provided a sounding board for ideas and concepts. Fred does far more than sell machinery; he is one of the industries forward thinkers.

Sterling Platt, now retired after forty-plus years of leadership and innovation, deserves special thanks. He still consults on occasion, but his absence has created a void in the industry. His son John is now stepping in to fill that void.

The book will likely create thoughtful discussion, and even some controversy. And if it doesn't, the author will be disappointed. However, there are some ground rules as the reader tackles the text.

It's all right to disagree with the author, but it is not all right to fail to offer constructive alternatives. Letters, E-mails (oakcreek@pond.net), phone calls, and other communications will be gladly received and a response will be made as appropriate. In this age of instant communication, the author encourages the free flow of ideas.

If a topic is particularly compelling and you desire more information, feel free to contact the contributors mentioned within the text or bibliography. I'm certain that they will demonstrate the same cooperation the author received.

The book is intended to reflect the work and thoughts of the author as well as the global participants within the forest products industry. A special thanks for all they do to respond to the needs of the billions of consumers worldwide. Forest industry workers, leaders, and professionals are an unselfish group overall. What other industry works today in nurturing a legacy that will only benefit future generations? It's that kind of industry, with that kind of people.

Dick Baldwin
Eugene, Oregon
August 15, 1999

Section One:

An Overview

1 The Final Decades of the 20th Century: A Springboard for the New Century

Historically, the forest products industry has had a simple, basic structure: finding the trees, making the product, and supplying the customer. As long as the trees were cheap enough, the mill operated in a reasonably efficient way, and there was a decent market, the mill made a profit. When there was no profit, the mill shut down.

As one century draws to a close and another begins, the industry is coping with unprecedented changes. Business norms of past years do not work very well anymore. The era of extractive forestry is ending, as environmentally conscious people have become involved and a greater number of laws and regulations are encumbering the landowner and producer. Recognizing and accepting those changes offers a springboard for the new century. If we want to actively shape the future—rather than defensively responding to these changes—we must understand the watershed events of the century's final decades (1970–2000), recognize how each has contributed to change, and restructure our practices.

The historic timber chase across North America became a global chase that is now also coming to an end. Fortunately for the forest—and for the future wood products customer—the public demand for something better, as well as the changing world economy, demand that the forest industry reinvent itself.

Timber and timberland have been too cheap for too long. In 1905, Gifford Pinchot, the nation's first Chief Forester, said that ". . . an acre of growing natural forest can be bought in nearly every forested part of our country for less than it would cost to plant." (Pinchot 1905) And where the forest wasn't cheaper, the industry moved on.

Yet even during those early years there were those who saw beyond extractive forestry. Actions in state legislatures began to demonstrate the power of an aroused citizenry. In 1903 Oregon and Washington passed their first forestry legislation, and supplemented it periodically in the following years. Michigan went even further, beginning a program of forest restoration in 1900 by taking over abandoned cut-over lands or burned-over lands foreclosed on by the counties for delinquent taxes, then replanting and consolidating these tracts and implementing active forest management. Other states, individuals, and companies have followed Michigan's early example.

Unfortunately, later in the century, depleted private holdings and lengthy harvest rotations forced the industry to rely increasingly on federal timber. Federal timber sales doubled during World War II and continued an upward climb to 6.2 billion feet in 1955.

With the exception of Weyerhaeuser, all of the forest product manufacturers in the west were, to some extent, dependent on federal offerings, especially in the Pacific Northwest. This situation could not last. Overcutting of the resource—with growth-to-cut ratios of only 60% or less in some areas during the 1970s, and unsightly massive clear-cuts—signaled the coming demise of extractive forestry. Space-age communication technology has brought to the public both the statistics and the disturbing images of such practices.

The Media at Work

Ralph Nader knows how to use the media. His choice of words, selection of topics, and careful timing of his approaches to the press have become familiar trademarks. Nader was an early critic of the industry, in 1974 charging that the U.S. Forest Service had long since abandoned conservation in favor of turning the national forests into timber factories.

Nader's remarks, along with others, prompted the forest industry to action. They realized that balancing the product needs of the consumer with the environmental expectations of the world public would be a tough task. Noisy minorities such as Earth First! began to play to the press in much the same way that antiwar protesters did during the Vietnam War. Outrageous staged events made the front pages and the evening TV newscasts. Coupled with the increasingly activist stance of some older, more traditional interest groups, the protests tapped into the public mood and captured the public's interest.

The public wanted something more from the forest industry than paper and wood products. An increasingly urbanized population wanted its forests, its open spaces, and its native animals preserved. Furthermore, the public wanted a say in how these values were obtained; first on public lands, and later on private lands.

This emerging trend was increasingly reflected in the media. Dave Pease, then Executive Editor of *Forest Industries* magazine, cited the lack of objectivity in the press reporting:

> Read now the lead paragraph of a UPI dispatch which appeared in our local afternoon daily Feb. 13 [1980], and you'll see the reason for my disillusionment:
>
> > Washington (UPI)—Interior Secretary Cecil Andrus has taken another major step to insure virgin Alaska lands will not be ravaged by man in the event Congress this year fails to pass measures to preserve them in national parks.
>
> It's hardly my role to attempt a re-education of wire service news writers, but here's a suggestion: If any of you happen to have a UPI bureau chief as a lodge brother, golfing partner, hunting or fishing companion, or drinking buddy—kick some ass. You'll be glad you did. (Pease 1980)

Forest and logs, the use of each will change with as the 21st century dawns.

Despite Pease's entreaty, in the following years UPI's coverage and the tenor of other international, national, and local newswriting continued much the same—a continuing source of irritation to the industry. Ad hoc industry-sponsored organizations formed, aiming to better inform the public. They distributed factual visual aids to schools, produced full-color descriptive brochures, and assigned public affairs personnel as press and community liaisons. Even the operating personnel communicated the industry's story. Forest industry managers broadcast pleas for understanding to state legislators, congressmen, and Forest Service senior officials. The traditional arguments for "logs and jobs" just didn't work anymore.

A letter to an editor in early 1997 sums up the frustrations of those who had failed to recognize the public's changing attitude about the forest and the wood products industry;

> The increasingly environmentally sensitive public [has] been subjected to years of media obfuscation, but no facts. . . . The need to communicate the basic, optimistic facts about the forest industry has been with us for many years, but it is nearly impossible to obtain media coverage of the information, or to reshape the educational process to present truth to our future citizens, or to combat the powerful propaganda of all the "green" organizations. (Rethinking the Industry 1997)

The fact is, the press may have recognized a trend in the public's thinking and then reinforced that trend. Jean Mater, in her insightful 1997 book *Reinventing the Forest*

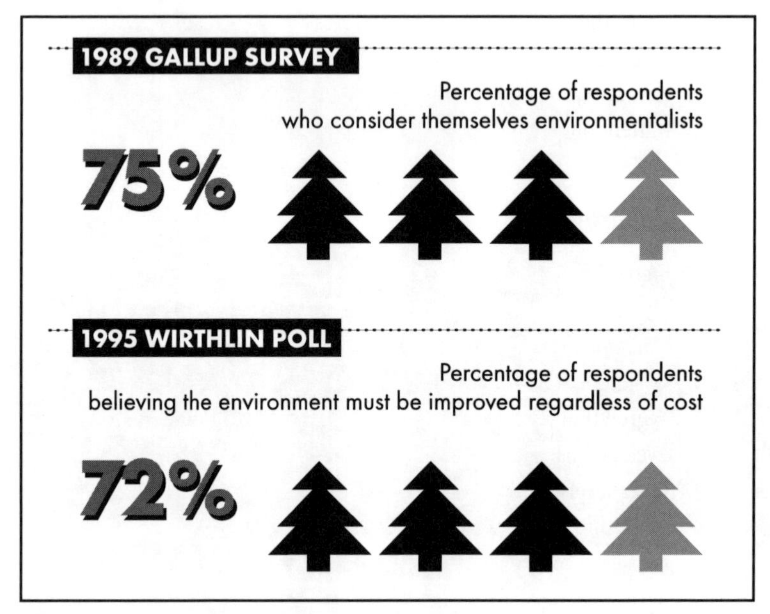

Exhibit 1.1: Environmental Leanings of the American Public

Industry, reminded the reader of two basic statistics. First, a 1989 Gallup survey showed that 75% of the respondents considered themselves environmentalists. Second, 72% of the respondents to a 1995 Wirthlin poll stated that protecting the environment is so important that improvements must be made regardless of the cost. (See Exhibit 1.1)

Those in the forest industry who acknowledge the realities of the emerging public sentiment, and use that information to proactively restructure their business, will increase their chances for survival. They recognize that the general public has a keen interest and an increasing voice in forestry practices. However, others still believe that the industry is simple and basic in structure, and should operate within its own environment. They still expect to turn a profit by simply finding cheap timber, extracting it from the forest, and taking it to market as a log or product. Many still cling to the belief that the public should have little say in how the forest is managed, particularly on private lands. But the public will have a say, whether officially or unofficially, as countless examples illustrate. Consider this one, recently played out in post-apartheid South Africa.

A sawmill owner near East London managed to become self-sufficient by growing vast tracts of pine timber on the southeast corner of the African continent, on the fertile lands between the escarpment and the sea. However, by North American standards the sawmill floor was overstaffed well beyond the needs of the logs, equipment, and rate of production. When industry observers questioned this, the owner replied that there is less likelihood of a fire starting in the pine plantations if the people of the nearby village share in the benefits of the forests.

In other areas of the world, the public is less likely to use fire to express its disapproval, or resentment at being left out. More often, the industry feels the public response in the form of new laws, regulations, litigation, consumer boycotts, public demonstrations, and other attention-getting activities.

Crisis or Opportunity?

The dual appetite for wood products and other forest values could well emerge as a resource-allocation crisis of epic proportions. For example, annual American per capita consumption of timber products climbed from 61.1 cubic feet in 1970 to 79.5 cubic feet in 1988, a 30% increase overall, while the nation's population rose just 20%. The net effect was a 55% growth in the nation's timber product consumption.

The other developed nations—and the developing nations too—show this trend. As people reach higher levels of literacy and education, they consume more wood-based products. Use of printing and writing paper, even though not all are made of wood fiber, is an indicator of overall consumption of forest-based products. Worldwide production of printing and writing papers increased nearly 84% between 1977 and 1988, and global production soared in the following years, more than doubling between 1977 and 1994. Mainland China's production and consumption grew even more rapidly, with an astonishing fivefold increase during the same period. In the face of such growing demands on a diminishing working forestland base, the working forest must work harder. In the United States, it is; although total forestland declined from 754 million acres in 1970 to about 731 million in 1987, the lands were more productive—as the 694 billion cubic foot of total growing stock became 766 billion cubic feet in 1987. However, the commercial harvest volume from the national forest system increased slightly during the same years, then plunged to less than half the 1988 volume during the following five years.

The western "owl wars" stemming from the Endangered Species Act, and the downsizing and redirection of the Forest Service, are blamed for the decline. Private lands in the United States and other countries now must make up the difference. In New Zealand, too, working forests have shifted to the private sector.

Of its 7.4 million hectares of forest—27% of New Zealand's total acreage—about 6.1 million hectares are natural forest. Seventy-nine percent of this forest is owned by the Crown, and nearly all of the Crown forest is protected as a conservation estate where logging is illegal. The remaining forest lands—plantations stocked with radiata pine and some Douglas fir—are intensively managed. As additional lands are placed in timber production, managers plant genetically improved growing stock, thus ensuring that growth of logs and wood products will continue to outpace the internal demand.

Chile, another heavily forested nation in the Southern Hemisphere, operates its forests and forest-based industry in a fashion similar to New Zealand. Although some harvesting continues within the indigenous forest, there is growing opposition to further logging or even active management of the native hardwoods. The Trillium Corporation's Rio Condor project is a case in point.

A vast, relatively untouched hardwood resource at the southern tip of the continent, covering about 275,000 hectares in Chile and 75,000 in Argentina, the Rio Condor forest is 80% linga with other hardwood species. As part of an arrangement with Chile and Argentina, the Trillium Corporation planned to establish wood-processing complexes in each of the two nations. The company proposed a carefully balanced mix of management practices to address social, environmental, and economic issues. Published stewardship principles were the operating bible. The goal was to attain world recognition as a producer of ecological sustainable products. The company earmarked a $200 million-plus investment package to provide jobs for about 800 employees. However, well-organized opposition disagreed with the company's goal, preferring wild lands to a managed forest. The Supreme Court of Chile subsequently ruled in favor of the environmentalists. Whatever the ultimate disposition of these lands, it is likely that more, rather than fewer, environmental constraints will result.

Chile continues to grow its forests and build its forest products industry through active plantation development, plus the installation of world-class conversion facilities that ensure efficient use of the abundant wood resource. A similar approach can be seen worldwide, as countries rely less and less on government forests or natural forests for raw material.

As the industry searches for alternatives to its traditional sources, three top choices have emerged: (1) supplying a country's needs through intensive management of the remaining lands available for a working forest; (2) recycling wood and wood fiber; and (3) finding substitute raw materials. Private landowners like Bob and Margaret Kintigh of Mountain Home Ranch will play a key role in meeting future log demand.

This western Oregon tree farm was planted in 1938 from cut-over tracts. Douglas fir plantations were thinned from below to 202 trees per acre at age 35. Two additional thinnings—in 1987 to 122 trees per acre and in 1994 to 78 trees per acre—resulted in increased growth in the remaining stand, as well as income to the owner. Bob Kintigh noted that if this plot had been harvested at age 44 (common practice), total yield would have been 34,240 (board measure). By maintaining the plot through fourteen more growing seasons, he now has a yield of 60,410.

Skid roads are narrow and few at the Mountain Home tree farm. Cutthroat trout are abundant in a nearby stream. The forest canopy allows some light to nourish the browse and wildlife hiding places. The Kintigh regimen of thinning from below results in the smaller-diameter timber being cut periodically during the life of the stand. The competition for light results in a straight stem; a higher ring count per diameter inch as the tree competes with its more dominant neighbors; and small tight knots as the limbs are naturally pruned. The resulting tree is ideal for Machine Stress Rated (MSR) lumber or Laminated Veneer Lumber (LVL) type veneer. A nearby nursery provides genetically improved replanting stock; thus, the yield improves during each forest life cycle.

A few small, inefficient sawmills are still operating as relics of the past.

Ongoing record-keeping aids in making improved management decisions. This productive industrial forest is an example of what has happened in Oregon and other states to meet the nation's growing wood needs.

Other countries and regions have provided similar examples for centuries, but now the need for additional large-scale projects is more urgent. We must retain disappearing tree species, such as teak, whose natural stands face extinction.

Teak grows only within a very narrow band around the world, within ten degrees of the equator. The last extensive teak stands may no longer exist, as warring factions in Myanmar use the country's rainforest to fund their civil war. Countries around the world are banning rainforest-derived timber because of growing ecological concerns. Allowing the rainforest to reestablish the species through protective set-asides is one solution; another is plantation forestry.

Flor y Fauna S.A., a Dutch-financed company headquartered in Costa Rica, began establishing a teak plantation on pasturelands north of the country's central valley in 1989. The climate, the enriched soils characteristic of the region, and a genetically improved teak species are expected to produce a final harvest in twenty years. The initial planting of 1,100 trees per hectare was precommercially thinned in half by year two. Beginning at year six, periodic thinning provides small-diameter cuttings for furniture parts and related products. As the trees mature, the higher proportion of lighter-colored sapwood—an inherent characteristic of the earlier cuttings—will rapidly give way to the more desirable darker hues of the heartwood. A nearby factory, specifically designed for small-diameter timber, provides income to the company—and rapid information back to the forester. The factory was designed and the equipment selected to support manufacturing changes as the teak matures. After mature harvest, the plantation is replanted. In time, this and other plantations will provide desirable tree species that are rare in commercial quantities outside the rainforest.

Designing the Tree

The Kintigh model for growing trees, the innovative teak plantations in Costa Rica—these illustrate the industry's emerging capability to *design the tree for the product.* Increasingly, the land manager can control the selection of species, genetic characteristics, soil conditions, harvest cycles, and growing conditions in general—designing the tree for the product, with both traditional and nontraditional wood products species. *Making* the tree—rather than finding it—will be the operating norm in the years ahead. One Central American plantation offers a further glimpse of this future.

Until recently, few thought the gmelina had much commercial value because it was difficult to dry. When planted in rich tropical soil it grows very rapidly—8 meters in height per year on a local soil-site index one. At harvest age 6, about 300 cubic meters of useable fiber are available for chips or pallet stock, but that may not be its best use. Root stock developed from stump sprouts of selected trees is yielding genetic characteristics that one day may result in use of this species for higher-valued shop and furniture cuttings. A 10-year-old tree in Panama may yet be an acceptable substitute for a century-old white fir of the western United States.

This type of research is important for the future requirements of the growing population. The highly visible environmental wars and controversies have distracted the forest industry participant, and sapped a huge amount of productive energy from the task of producing an ample array of wood-based products. The consumer's needs will be more urgent and more diverse: more urgent, because a growing world population is expecting more wood products per capita from a shrinking working forestland base; more diverse, because the same consumer who wants a wider array of products also wants that choice of products to be compatible with good environmental stewardship. Finally, the consumer wants special forested places restored and reserved. That's a tall order, one unlike any faced by earlier generations in the industry. Events of the late 20th century, and the industry's response to those events, must provide a springboard for industry growth and change in the new millennium.

2

Raw Material and Environmental Issues: A Changing Paradigm

A paradigm is a way of visualizing or thinking about the world, and like a habit, it is difficult to change. The long-held paradigm of the forest visualized logs as the primary product. The world was full of trees—they had always been there, they would always be there. We planted, we opened up the forest at intervals for harvest, and then we either replanted or abandoned the land. We consumed the products throughout our everyday life; forest growth and consumption, except for the setting aside of the special places, was accepted as normal.

As the decade of the 1970s drew to a close, there were telltale signs of a challenge to this traditional paradigm. Timber prices skyrocketed, housing starts fell, and the public contested the industry's right to log the land. How could this happen? Why weren't the traditional economic arguments of logs and jobs working anymore?

A Paradigm Shift

The forest industry failed to recognize that a shift was under way—a paradigm shift in which an increasingly urbanized population set aside long-standing values and thought processes. The forest industry labored to regain its stature and timber, while urban citizen coalitions challenged timber and logs as the primary forest benefit.

Challenging timber sales, litigating timber practices, and demonstrating against the cutting of timber became accepted ways to restrict access to timber. As public land use changed from tree growing to preserving sanctuaries for endangered species, harvesting on public lands came to a near standstill. The industry began to search in earnest for understanding, pitting facts and reason against beliefs and emotions. The search took years, with frustration mounting on all sides. Each year the available timber and logs from traditional sources declined further; debate and conflict continued. And what was happening in timber issues in the United States was also occurring worldwide. A paradigm shift was well under way in the world at large, as world-renowned futurists and authors Alvin and Heidi Toffler wrote in 1994:

> A new civilization is emerging in our lives. . . . The Second Wave—the rise of industrial civilization—took a mere three hundred years. Today history is even

> more accelerative.... Those of us who happen to share the planet at this explosive moment will therefore feel the full impact of the Third Wave in our own lifetimes. (Toffler 1994)

What is the Third Wave? What does it mean to us? What does it mean to the forest industry?

> The Third Wave brings with it a genuinely new way of life based on diversified, renewable energy sources; on methods of production that make most factory assembly lines obsolete; on new, non-nuclear families; on a novel institution that might be called the "electronic cottage"; and on radically changed schools and corporations of the future. The emergent civilization writes a new code of behavior for us and carries us beyond standardization, synchronization and centralization, beyond the concentration of energy, money and power.
>
> This new civilization has its own distinctive world outlook; its own ways of dealing with time, space, logic and causality. And, its own principles for the politics of the future. (Toffler 1994)

Nowhere are the new world outlook, new code of behavior, and the "own principles for the future" more evident than in the forest industry. Dr. Jean Mater, a respected industry figure, later stated:

> As environmental awareness emerged in the last quarter of the twentieth century, social responsibility promises to be the cornerstone of the twenty-first century. The new definition of any institution's role in the traditional mission—the production of goods and services to meet the needs and wants of customers. Heretofore, customers weighed their needs and wants against price and quality. Today, many consumers add environmental and social acceptability to the attributes they look for in products. (Mater 1997)

Chapter 1 touched on the changing public perception of environmental issues. The linkage between environmental acceptability and perceived social responsibility has become an integral part of the buying decision. Surveys and polling results during a recent seven-year period all confirm the Toffler theory that defines the basic elements of the Third Wave—the information age.

The consumer not only worries about the loss of natural habitat, but also believes that protection of the environment should be given priority over economic development. If there is a choice between the survival of a rural community and of an ancient forest, the forest comes first. In contrast to the frontier ethic of individual self-reliance, 77% believe that government regulation makes the environment a much cleaner and safer place (Mater, June 1997), and 62% want Congress to make environmental protection a priority over timber cutting (Mater 1997).

A 1992 survey revealed that two-thirds of those polled will buy environmentally sound products and 83% make an effort to reduce energy (Mater, June 1997). The

Rural timber-dependent communities have become an endangered species.

conclusions are transparent—not only has the industry been using logic to confront belief and emotion, it did not recognize a changing paradigm that departs from the industrial-based economic view of the world. Recognition of this emerging view, which sees and thinks about the forest as part of a larger world experience, can give the industry a basis for going forward into the new century.

The Power of Words in Defining the Paradigm Shift

Words facilitate the acquisition of knowledge and understanding. When people communicate, words are the key. Sometimes words convey facts; sometimes they convey opinions. Sometimes they excite emotions; sometimes these emotions overshadow and overwhelm the message. A well-understood meaning, agreed upon by the communicating parties, is essential to mutual understanding.

It is common for words to be used with the assumption that their definitions are understood by all. But this assumption frequently translates into misunderstanding. For example, the word "biodiversity" conveys multiple meanings and conflicting images—this term, and its longer source "biological diversity," have been defined in at least 85 distinct ways (Adair). How would you define the term "cumulative effects"—or even the word "sustainable"? Try the expression "ecosystem management." Now describe its "biophysical" and "socioeconomic" dimensions. Further define its cousin, "ecological complexity" and then try forming a mental image of a forest's "spiritual quality."

When far-reaching decisions are documented with words that share several meanings—without clearly defining those meanings—there is grave danger of misunderstanding. The word "ecosystem" offers a prime example.

> The U.S. Fish and Wildlife Service distributes a map showing 52 defined ecosystems for the United States, Puerto Rico and the U.S. Virgin Islands. The U.S. Forest Service and U.S. Dept. of Agriculture do the same. None of the maps are the same! Huge differences in patterns of ecosystems exist showing that an ecosystem truly exists only in the eye of the beholder! (Adair)

If the discussion is about ecosystems, and the participants can't agree upon the definition of the word, it is unlikely that they will agree when lines are drawn on a map.

The various stakeholders must insist on agreed-to definitions prior to entering discussions that will lead to decisions on forestry and associated environmental issues. This may require a preamble to new documents and a reworking of existing agreements, not unlike the format customarily used in binding legal contracts.

After hammering out the definitions, the parties then move on to determine the likely outcomes. The notion that the good guy agrees and the bad guy disagrees on expected outcomes will only resurrect the barriers to understanding that were dismantled by the agreement on word definitions. Agreement on language and words provides a tool for understanding—the process can then be either expedited or hampered by a good guy/bad guy perception of the other players. Using words to facilitate communication and understanding, the parties must willingly collaborate to reach solutions on forest-related issues.

The Changing Role of the Forest Manager

The science of forestry and forest management is of comparatively recent origin; its origins date back to the late 18th century. The writings of Hartig and Cotta, two German foresters, provide the foundation for modern forestry. Their influence soon crossed national boundaries, as neighboring countries such as France, Austria, Sweden, and Switzerland adopted the developing forestry science.

In Switzerland, the Canton of Bern issued an academic degree for protection of the forest as early as 1592. However, the nation didn't respond until a federal forest inspector was appointed in 1875. In 1876, a new law mandated forest practices on slopes to protect the lowlands against floods, avalanches, and other similar dangers of wind and weather. Elias Landolt, the first Swiss forester, later said:

> Our forest laws are intended to work more through instruction, good example, and encouragement than by severe regulation. This method is somewhat slower than one which should involve harsher measures, but the results achieved are

Big logs represent the past.

more useful and lasting. When forest owners do something because they are convinced of its usefulness it is done well and with an eye to the future, but what they do under compulsion is done carelessly and neglected at the first opportunity. What they have come to learn in this way and have recognized as good will be carried out, and that better and better from year to year. (Pinchot 1905)

The Swiss forestry laws embodied bold legislative concepts that we are still grappling with today. They stated that the forest is important to all, and that the replantation of trees must follow their harvest—early ideas that have yet to find universal acceptance. An essential part of the early legislative framework was the concept of citizens placing forestry restrictions on themselves for the benefit of the commonwealth, and the notion that instruction, good example, and encouragement are more effective than severe regulations.

Swiss and German forestry concepts were well understood by the first U.S. Chief Forester, Gifford Pinchot. He literally wrote the book on how to manage the American forest reserves that were the forerunner of the present-day national forests. He defined and prioritized the tasks to administer over 63.3 million acres of land: first, to protect watersheds and drainage basins; second, to supply grass and other forage for grazing animals; and third, to furnish a permanent supply of wood for the use of settlers, miners, lumbermen, and other citizens.

The original forest manager was the forest ranger who made decisions as needed. The ranger lived in the community and benefited from that close association. The forest ranger had a lot to say about the forest, and some say about meeting the needs of the public. The task was pretty basic—plant trees, nurture and protect them, and make certain that the other values such as graze and watershed were protected. These tasks were the basics of forest policy, a policy grounded in the defined needs of a rural, resource-based economy.

Today, the task of growing trees and meeting society's expectations is far more complex for the forester and the forest manager. However, the basics remain the same: define the expectations of the owners and customers, develop a policy that articulates the structure to meet those expectations, and move forward in a proactive way. To meet the demands of an increasingly urbanized population takes more than the usual complement of technical skills. It requires vast changes—essentially, a renaissance in forest policy.

The roadblocks are daunting on a regional, national, and international scale. The modern forest manager role requires a host of skills, political as well as technical. The "new" forest manager will:

- Define and understand the forest issue to be decided. Is it a land use question? Is it a matter of harvesting practices? Is it an environmental problem?
- Recruit and enlist individuals that represent a broad segment of the user groups, with certain knowledge or experience; then, structure meetings or communication processes so that each representative has equal voice.
- Use a structured, inclusive, and collaborative process to balance a sometimes overzealous moral authority claimed by interest groups.

- Get the assemblage to reach accord on what individual words mean and how each will be used.
- Work with the involved individuals to determine what method to use to obtain the outcome; that is, does the majority rule and are there winners or losers, or do the participants want to reach a consensus for the stakeholders?
- Clearly define goals, and set structure and deadlines for the work product. Schedule the collaborative process, and provide the organization and administrative leadership to create an efficient process.
- Provide the needed information: facts based on real science when available, and factual observations or experience of others as appropriate. Recognize that opinions, beliefs, and causal observations may have little value when dealing with complex situations governed by natural law.

The forest manager increasingly relies on the various owner and nonowner constituents for direction in managing the forest. The Gifford Pinchot-type forest ranger has given way to the decision-making of involved stakeholders—or groups that fashion themselves as stakeholders. The new realities demand a different approach. The forest manager and landowner really have little choice but to understand and adapt.

3

Research and Development: Making It Work

The demand for technological change in the opening decades of the new millennium is greater than at any time in the past century. Growing consumer expectations, a shrinking working forestland base, and a host of other issues all create a compelling need for technological development in the forest and at the mill.

Ross S. Whaley, President of the College of Environmental Science and Forestry at State University of New York, said it best:

> While forest industry success in the past two decades has been determined mostly by marketing, financial, and political skill, in the future, success will be influenced to a greater extent by technological adaptation . . . the key to solving many of the environmental concerns facing forestry and the forest products industry will be improving technology, not limiting technology, which is often proposed by some well-intentioned environmentalists. (Whaley, March 1996)

What is the current research? Who is conducting it, and how? What can the industry expect? Which companies are doing it right? What resources are now being committed to the task of research and development (R&D)? The need for answers to these and other questions grows more urgent.

Funding Research and Development

How then is the forest industry doing in committing resources to the task? The dollars tell the tale. The total annual public and private investment in forestry and forest products R&D in the United States was recently estimated to be about $1.3 billion. However, the value of shipments for the industry—logging to lumber to paperboard products—is well in excess of $210 billion. That means that for every $100 in sales, only about $0.62 is spent on R&D.

Already low in comparison to other industries, only about 6% to 8% of the research dollar goes to developing new knowledge. The remainder goes toward technical development, including finding the answers to environmental problems attributed to growing trees and making products. A 1993 National Science Foundation study identified these facts: forest industry expenditures per research scientist were below those of the petroleum, transportation, chemicals, primary metals, and food sectors. Only the fabricated metal industry spent less than paper and allied products companies. R&D in the state of Oregon illustrates the situation.

The Oregon-based forest industry generates an annual direct payroll of about $2.0 billion and, indirectly, an additional $3.0 billion in related manufacturing, transportation, and services. Tourism and outdoor recreation, much of it centered around Oregon's forests and related attractions, generates more than $3.0 billion annually. However, the Forest Research Laboratory at Oregon State University, a world-renowned R&D research institution, gets by on annual expenditures of about $9.0 million per year. This equates to about a penny per $100 of benefits to Oregonians. Seventy-six percent of the lab's income is derived from grants and contracts, a large portion of this from out-of-state sources.

The lab's productivity is incredible: a total of 264 forestry-related publications, issued between July 1, 1994 and June 30, 1996, document the research on forest regeneration, forest ecology, protection of forests and watersheds, forest use and practices, wood processing, and product performance. The Oregon lab is a bargain by any measure. Other labs, such as the research facility at Mississippi State University and at least nine more labs in the United States, perform comparably.

Research and Development Results

The New Zealand Forest Research Institute at Rotorua typifies what is happening in the Southern Hemisphere. This laboratory on the North Island conducts research and development with admirable cost/benefit ratios; a well-defined mission statement guides its efforts. Its funds come from its customers, including the government and private sources. The New Zealand activities range from "DNA fingerprinting" of trees to biosecurity, an integrated forest protection program that addresses risk, investigates new threats to the forest, and mitigates existing problems. The lab works closely with industry clients through a Forest Health Collaborative. It also functions in another unique fashion, acting as an incubator for new businesses under the Technology Commercialization section. For example, Greenweld Technologies Limited (GTL) teamed up with an American manufacturer to market a technology for finger-jointing green lumber. Short green lumber lengths can be finger-jointed and bonded into more desirable lengths. This results in higher lumber recovery from the log and a more predictable length assortment.

Other current startup businesses nurtured by the New Zealand institute range from Wood Hardening Technologies Limited to Lektraspray, producing an electrostatic spray system for application of anti-sap stain chemicals. Various software programs are also being commercialized. A wholly owned subsidiary, FRI International Limited, markets and sells research management and technical expertise internationally. Projects in Turkey, Sarawak, and other developing communities have benefited from this New Zealand research initiative.

The spindless lathe is an example of R&D meeting changing needs.

The Oregon and New Zealand labs are examples of well-run businesses using commitment and ingenuity. The New Zealand lab is governed by a board of directors and staffed by a mix of professionals. Everyone, including board members, is compensated for their services. A full-time chief executive provides leadership for 11 research disciplines. The annual report and the accompanying financials leave no doubt that the lab is operated as a for-profit business. The business is even taxed, yet it had an after-tax surplus of slightly over US$1.0 million in 1997.

The Oregon lab is somewhat different, being affiliated with a university. The dean of the college of forestry is the administrative head; a volunteer board of industry professionals from the public and private sectors meet periodically to review present programs, review areas of research, and provide recommendations for the future.

The Forest Research Institute of Malaysia, Forintek in Canada, and labs in Japan, Europe, North America, and other regions of the world also contribute to the forest industry knowledge base. Company labs, sponsored by manufacturers and suppliers, also assist the R&D effort. Most make contributions, but there is a common thread separating the really good from the also-rans.

The good labs have well-defined mission statements or charters. They are organized for results, and a willing customer pays a fair toll. Keeping the process open to outsiders ensures continuous feedback and input. Conversely, those laboratories that provide a service that is not well defined, that have weak leadership, and that are reluctant to share their results with others usually have little to offer their constituency or the industry at large. One industry reporter made this observation:

> We've got the expertise . . . [they] are experts in their fields; they are sitting on a wealth of information that could be of immense benefit, but for whatever reasons, they aren't disseminating it. That has to change, because we've just about used up all our past glories. It's time to rev up the old technology transfer machinery again and start making some of that knowledge and expertise available. (Johnson 1997)

Today's research needs are more urgent then ever. The past emphasis on technical expertise in logging and process has shifted to learning how to grow more and better fiber—and grow it more quickly, in an environmentally prudent manner, while coping with a smaller log and a more critical public. That is a tall order by any measure.

The Tale of Three Companies

Research and development is like a treadmill; you mustn't stop or you will fall behind. Sir James Goldsmith said it differently, but said it well: "If you just do the same as everybody else, you have to fail. It's inevitable. There is no escape . . . " (Wansell 1987) The successful forest industry companies are dedicated to doing something different that also provides a measurable competitive edge. Three case studies illustrate the range of company experiences. The first, Trus Joist International (TJI), is a case study in doing it right big-time.

Trus Joist International (TJI)

It's no secret that investors love a high-tech stock—and a successful high-technology forest-products firm such as TJI. Growing customer acceptance of the products reinforces that appeal. Harold Thomas and Art Troutner, the company founders, knew they had a superior wood product when Laminated Veneer Lumber (LVL) came to market during the late 1960s. They probably were not thinking about "high tech" at the time. They were concerned about the scarcity and high cost of lumber that had the "just right" structural properties. Troutner's early research and development work provided solutions to the lumber problem. They found a way to just manufacture lumber using a plywood-like process: peel the log, dry and test the resulting veneer for structural attributes, then glue, assemble, and hot-press the veneers into a billet.

In the 1970s and 1980s the company grew and prospered. Technology and equipment patents identified the manufacturing steps to produce a lumber billet two to four feet wide and up to 68 feet in length. Flanges for I-joists plus headers, beams, scaffold planking, and other lumber products could be remanufactured from the billet of parallel laminated veneer. Large log products could now be manufactured from logs as small as eight inches in diameter.

For example, the I beam was constructed with an Oriented Strand Board (OSB) or plywood web of varying thickness and depth; the flanges were precut strips of LVL. The width of the top and bottom flange, the thickness of the billet, the mix of veneers, and the thickness and height of the webbing material provided the desired properties for the I-joist. The I beams made by TJI came in a variety of sizes and structural characteristics. They weighed less than the solid lumber that they replaced, were more dimensionally stable in the field, and exhibited more consistent structural properties. The product was easier to handle at the job site than solid wood or metal options.

Research and development continued at the lab and in the mill as LVL became a commercial reality. The evolution of the idea took a quantum leap forward in 1991 when Trus Joist joined forces with the Canadian giant MacMillan Bloedel (MB). The new company pooled the Trus Joist technology with pioneering MB engineered wood technology to form the industry's foremost high-tech company, Trus Joist MacMillan. The jointly owned company combines the resources of both companies with the visionary leadership of the Trus Joist organization.

Trus Joist, the controlling entity of the TJM partnership, soon became Trus Joist International (TJI) in recognition of its emerging international stature. TJI sales revenues of $66.0 MM in the 1982 recession year expanded over tenfold in the following 15 years, to $700.0 MM in 1997, as market acceptance of the technologies continued to grow at an exponential pace. The trend continues, with recent year-to-year sales growth of more than 22%.

The market price for TJI shares has followed its successes. The TJI common stock went from $17 per share in 1995 increased to $33 in the first quarter of 1998. This breakout is uncharacteristic of the other forest products stocks; most have done little over the same time period. For TJI, the best may be yet to come.

MacMillan Bloedel

The wood products side of MacMillan Bloedel, the second case study, features the usual mix of structural panels, lumber, and a variety of other products. The paper sector is equally unexciting: newsprint, linerboard, and related paper products make up the bulk of its product mix. The company's extensive forestlands, well over 4.5 million acres, provide cash-flow stability. The crown jewel that made MacMillan Bloedel stand out was its R&D resources. The company's R&D efforts created a new generation of engineered wood products such as Parallam, a lumber product composed of parallel laminated strands of wood milled from dry veneer. This product—an engineered wood solution for the solid wood timber and beams historically used in big timber construction—grew out of MacMillan Bloedel's laboratory efforts. So did laminated strand lumber, which is made by taking pulpwood logs, some as small as 5 1/2 inches in diameter, and milling each piece into strands up to 13 inches in length. The strands are then compressed under heat and pressure into a board up to 5 1/2 inches in thickness. Header and rim boards and other uses are emerging.

MacMillan Bloedel continues to benefit from this new technology, but only as a minority partner with TJI in the Trus Joist MacMillan (TJM) partnership. The parent company, MacMillan Bloedel, still relies on its larger mix of commodity products. This continued reliance on an aging product mix—and the company's inability to use new technology to its advantage—have resulted in declining sales and earnings as the century draws to a close. MacMillan Bloedel's 1997 earning per share was a negative $2.99 Canadian, with the share price hovering in the $13 to $14 range before plunging to less than $11 at year-end. MacMillan Bloedel then recruited a new chief executive, Tom Stephens, to breathe new life into the company. This new life included pulling out of many of the pure commodity businesses, downsizing the company in general, and closing its research and technology facility in Burnaby, British Columbia. The company will survive, but the question lingers: Will it get a second chance to achieve world-class status? Or will it effect a strategic merger with another company? Recently, Weyerhauser Company bought MacMillan Bloedel.

Louisiana Pacific

Louisiana Pacific (LP), our third case study, is a Cinderella story. The company was a late 1960s offspring of a Securities and Exchange Commission (SEC) action pending against its parent, Georgia Pacific Corporation (GP). The SEC feared that GP's rapid growth and market dominance in southern pine plywood would result in a commanding dominance over the wood products industry, particularly in softwood plywood manufacture. Harry Merlo, a GP vice president, took the helm as chief executive of Louisiana Pacific. During the early years, commodity lumber, softwood plywood, and redwood lumber were the sales and marketing mainstays. But all that changed when Merlo discovered waferboard.

Waferboard was then being manufactured by a small handful of companies. Customer acceptance was lukewarm, production costs were high, and profits were meager in the best of times, but Merlo saw the possibilities. LP managed to improve the board properties by orienting layers of strands, much like the strength-enhancing characteristics of plywood. A cookie-cutter plant design evolved as plant after plant was brought on line. The resulting product, now called Oriented Strand Board (OSB), found new life as a more affordable substitute for softwood plywood. Louisiana Pacific's OSB production was 29.0 MM surface measure in 1978, increased to 95.0 MM in 1981, and leaped to 373.0 MM in 1983. LP had one production plant in 1978—and nine in 1985. Twelve more came on during the next decade. And that was only the beginning. LP added a host of other technologies, such as Laminated Veneer Lumber (LVL) and Medium Density Fiberboard (MDF). A variety of insulating board plants used a diverse mix of raw materials—one plant in the northeast relied on urban waste for its furnish. LP's industry-lookalike balance sheet was transformed into a Wall Street star. Annual net income before taxes of $57.0 MM in 1976 climbed to about $200.0 MM in 1988, then a major breakout occurred between 1993 and 1994, with a two-year total of nearly a billion dollars. But the Cinderella story was not to last.

Louisiana Pacific was well known as a lean and challenging competitor. Less well known was the fact that product manufacturing controls had not kept up with growth, and that an inadequate R&D effort supported the high-tech products being manufactured. The "new breed of company," as it billed itself in company literature, was selling problem products.

"OSB is a hell of a good product," said David Beaty, head of the building department of 500,000-resident Seminole County in central Florida, "but it's very susceptible to moisture . . ." (*The Business Journal*, December 16, 1994).

OSB was drawing a lot of complaints. One product, OSB-based lapsiding, was failing on hundreds and even thousands of homes. The rapidly escalating complaints mushroomed into a $375 MM class action suit. In addition, apparent lack of compliance with quality protocols at two OSB locations led to criminal charges. A board decision soon followed; Merlo and several senior managers were ousted by the board or terminated by the new management.

The first order of business for Mark Suwyn, Merlo's successor and a veteran industry senior manager, was to reinforce and implement additional quality assurance programs for each of the mills. He then installed a Vice President for Technology and Quality, and move forward to settle the various claims and legal issues. Whether Louisiana Pacific can reemerge as the envy of Wall Street is anyone's guess. Suwyn and his board are betting that it can be done.

Making Research and Development Work

These three companies provide insight into research and development as a vital tool for superior performance. Trus Joist managed it well, sought outside technology to complement the in-house developments, and remained committed to the process for the long term. MacMillan Bloedel's enviable R&D effort ran counter to the corporate culture—a culture that expressed an unwillingness to adapt and grow a high-tech business. Louisiana Pacific had a vision—a vision built on the technology of others—and a culture that forgot "who brung them to the dance."

The question confronting wood-based product companies is not whether to have an R&D function, but what form that R&D function should take. Will the function be in-house, an outside contract function, or access to the work of others? Will the company be an industry leader, or a fast follower? Will technology be used as an enabler of change, or will it be used to incrementally improve the present raw material, the company's products, and manufacturing technology? Is the company willing to break out of the box to achieve the kind of prosperity and performance enjoyed by Trus Joist MacMillan—and by MacMillan Bloedel and Louisiana Pacific early on? How does the company cope with periodic downturns; does it prepare for growth, or react defensively? Can the company and its leaders identify the early warning signs of emerging trends, or does it look for bandwagons that are usually too crowded to be profitable? These questions trigger interest and curiosity. Finding the right answers brings the most profitable benefits from an R&D effort.

The fundamental question is, what is your company and organization doing to prepare for what's ahead in the 21st century?

Section Two:

The Raw Material Base: Globalization and Re-engineering

Wood Products and the Search for New Directions

4

Wood products and the manufacturing technologies that produce them are evolving and changing. The 21st century will be an age of great consumer marvels and additional advances in both product and construction technology. How, then, should the wood products manufacturer define the road ahead? Where can the opportunities be found? This chapter and those that follow will provides facts, observations, and insights on a growing, changing, and increasingly global wood products industry.

Raw Material and Wood Products

Logs from traditional timbered areas of North America and other regions of the world still constitute a major supply of raw material. However, short-rotation plantations, wood-based urban waste, and other nontraditional sources will contribute a growing proportion of material. Tree crops of selected genetically superior species in both the Northern and Southern hemispheres are becoming wood factories of unprecedented scale, managed more like agricultural crops than traditional forests. Urban wood—such as waste paper, pallets, tree limbs, construction debris, and other wood fiber—will complement both the traditional forests and the new plantations.

The growing array of raw material from both rural and urban sources will include wood fiber and increasing volumes of other organic and nonorganic materials. Organic sources include agricultural byproducts—straw and crops like kenaf, which is grown specifically as a substitute for trees. Nonorganics like gypsum, cement, and mineral sources can be combined with wood fiber to yield a product with attributes combining the best of each. Even the use of the whole log continues to change. Species and diameter preferences are becoming less important to the producer. Whole logs end up peeled, flaked, wafered, or otherwise milled into smaller segments that are reconfigured into attractive and functional panels, lumber, and other products. Small sizes become more useful as the wood conversion process becomes increasingly oriented toward new technology and small trees.

Supplying Wood Products Markets: The Emerging Model

Wood product supply is both challenging and changing. The supply of commodity lumber, panel, and paper continues to surpass requirements, due to an array of raw material and conversion options and the ease of market entry. A vast global movement of goods will occur as producers identify and target undersupplied regional markets. Response to demand will speed up as tariff and other trade barriers fall. Raw material will come from an expanding variety of sources, with at least several options in each consuming region. The emerging wood product supply is identified in Exhibit 4.1. The three major product processes will be (1) reconstituted or other reconfigured fiber-based products, (2) the more conventional solid sawn wood, and (3) veneer-based products. A product could be the result of one process or several. Two or even three processes could be involved to produce a single product such as an I beam. The new product may yield the same attributes of older more conventional commodities, or be distinctly improved. Furniture offers a good example of changes occurring in the composition of a manufactured product.

It is increasingly difficult to assess the structure of a stylish piece of wood-based furniture. It can be a solid wood assembly, or it may be constructed of a low-value hardwood and Medium Density Fiberboard (MDF) with a well-disguised thin precious veneer or paper overlay. The composite piece may be more attractive and functional than its solid wood counterpart.

This lookalike example is not limited to furniture. A new generation of building systems, built around Engineered Wood Products (EWP), is emerging as the top choice of a growing number of architects, specifiers, and builders. Currently, EWP and EWP-based building systems have barely reached their adolescence. Price is rivaled by technical service, a feature included by most EWP suppliers, as a basis of choice in the marketplace.

Consumers, always price-driven, are becoming more educated in their search for value. The wood products manufacturer—whether producing EWP, bulk commodities, or niche products—must respond by seeking out profitable market segments. Low manufacturing costs and an established customer base have worked well in the past and will still be a major contributor in the future. However, these factors are of limited importance when competing in oversupplied markets that accept a growing variety of products with comparable visual and structural values.

The successful manufacturer will be more agile and mobile to obtain an optimum combination of raw material costs, direct manufacturing expenditures, and transportation economies. In fact, transportation expense and the speed and certainty of delivery are becoming major competitive factors. The last decades of the 20th century provide just a glimpse of the wood products transportation matrix of the future. The expanding worldwide industrial base for a product type, combined with a wider choice of raw materials and merchandise that can fill a specific need, interconnects with a worldwide network of ships and modern ports. The ports that move bulk and containerized

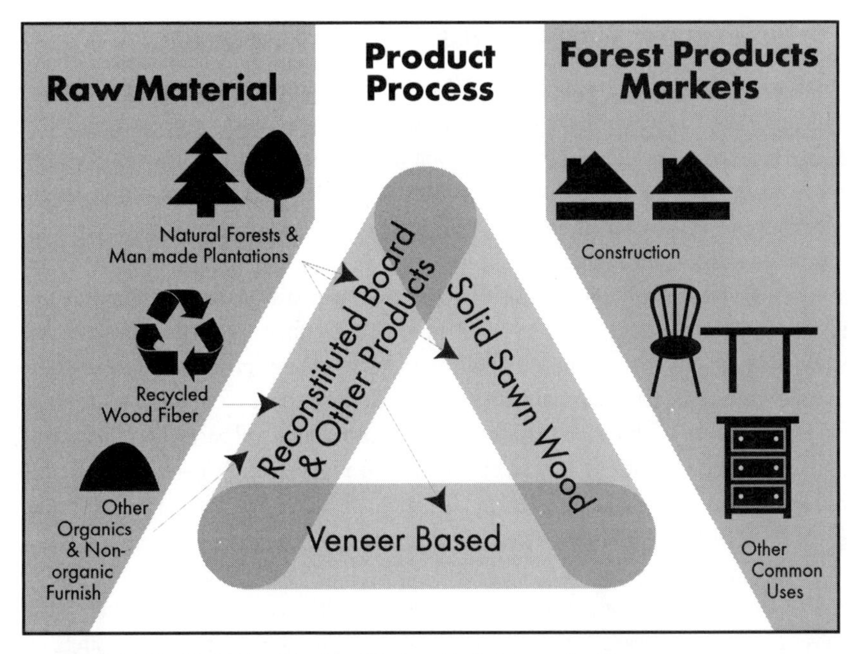

Exhibit 4.1: The Model: Supplying Wood Products Markets in the 21st Century

cargoes of food and other goods provide the framework for the movement of more wood-based raw materials and products. A land-based network cooperates with this transit service.

The capital funds sunk into constructing a manufacturing facility and the carrying cost of the capital employed will also loom as a major competitive consideration. The choice of product will often hinge on the installed cost of one process versus that of another. In addition, the original capital cost of a new process or manufacturing facility will be driven down as like facilities are constructed. Innovative and agile operators will design and build plants that combine the old and the new, yet yield manufacturing costs that compete favorably with better-funded competitors. In the new century, the greatest danger for the large, well-funded company may not be a larger, more technically advanced facility—it may be a smaller competitor using alternate technology and product with a closer-to-market facility.

The "big steel versus the nimble newcomer" paradigm, which played out in the 1970s and 1980s, may be the wood and paper paradigm of the new century. One thing is certain: having a cheap patch of timber and a modern, well-capitalized facility will not guarantee survival. The cycle of change and renewal for raw material, manufacturing facilities, and customer demand continues to accelerate into shorter time frames and tighter decision loops. A company's need for cash flow and a nation's need for foreign exchange play a role in determining who competes in which markets. The producer will be challenged to keep up with the pace.

Seeking Out Global Opportunities

The use of wood is growing nationally and globally. The most promising business opportunities in the new century differ markedly from those of the past, when producers chose both the site and the conversion options based on two essential factors: timber quality and the capability to transport logs to the mill. These resource-driven factors once overshadowed all other considerations. Now, however, the wood products industry is rapidly shifting from being resource driven to being market driven. The entrepreneur, faced with a burgeoning choice of raw materials, processes, and products, must consider the market's needs—and how to meet them profitably. Despite the importance of both the raw material source and the logistics of transport to market, customer requirements dominate the better global opportunities.

Manufacturers reaching out to the consumer face only one real barrier—that of self-imposed product limitations. It's not enough just to produce lumber—to succeed, the producer must satisfy customer demand . A successful example may be found in the new Acosa MDF plant in Ecuador.

Ecuador, with a population of 11.7 million and a 1996 GNP of $17.5 billion, has a small economy by any measure. The population is growing at about 2.2% annually, with an average annual increase of private consumption of goods and services per capita of 0.5% per year, as reported by World Bank sources. By comparison, the U.S. population is increasing at an annual rate of 1.0% and its consumption at a rate of 1.5%.

The Acosa MDF plant, located in the Ecuadorian highlands, represents a shift from the business founded on a natural forest base. Originally a tropical plywood plant, its later move into particleboard (PB) and plantation forestry set the stage for the new MDF facility. When raw material growth outpaced harvest and exceeded the PB plant's domestic demand, Acosa began searching for other conversion options. Betting that there would be a market in the developing Ecuadorian economy, the owners built a 55,000 M^3/year plant. Production, now about 32,000 M^3/year, is gradually growing. The formula for success? Strategic positioning of raw material, a desirable plant location for shipping, and choice of the right conversion option. To ensure continued growth in production, the owners have begun exporting MDF to nearby countries—and have installed an MDF customer training center.

The Acosa plant was fortunate in having raw material readily available. However, MDF and other wood-based products can be manufactured from alternate materials, including recycled fibers. Finding the market and manufacturing a product from available resources is the key. Some of the most promising business opportunities may appear in the most unlikely places.

Index of Economic Freedom

Growing urban areas within a region whose forestlands are being reduced may provide excellent plant location sites—as long as the country or region is business-friendly, the per capita wood products consumption and emerging demand are not being adequately met by existing facilities, and the manufacturer is clever enough to

A 12-year-old plantation log replacing rain-forest hardwoods, Limon, Costa Rica.

pick the right product and locate a sustainable and economical raw material source. Whether or not an opportunity exists depends on population size, annual growth and consumption trends, and the economic freedom to meet market expectations. However, these factors have limitations—they indicate emerging demand on a macro level, but miss micro demand.

Tables 4.1, 4.2, and 4.3 illustrate a comparison between nations, based on a list from the 1998 World Bank Atlas and the 1998 Index of Economic Freedom, identifying criteria of population, the growth rate of that population, and the annual consumption rate for a six-year period. The tables include nations that have combined annual population growth and consumption rates of 5% and more. Several nations with lower population rates but relatively high wood products consumption rates are included for comparison.

Each table lists nations with continuing or developing wood product market opportunities, divided into three population categories. Table 4.1 lists those with a population of over a hundred million. Those with a population between 21.0 and 100.0 million are found in Table 4.2; those with a population less than 21.0 million, in Table 4.3. (Note: An individual nation risk/reward appraisal is best obtained by an in-country assessment.)

The column titled "Index of Economic Freedom" is based on the criteria defined by the 1998 Index of Economic Freedom, published by The Heritage Foundation and *The Wall Street Journal.* Each country is ranked individually for each of ten economic freedom indicators, described below. The lower the overall index number, the better the opportunity (on a scale of 1 to 5). The ranking is limited to the current conditions; trends and individually negotiated arrangements are not considered.

- **Trade Policy.** The impact of tariff and nontariff barriers to the free movement of goods and services, with 1 indicating average tariff rates of 4% or less and/or very low nontariff barriers, and 5 indicating an average tariff rate of 20% or more and/or very high nontariff barriers.
- **Taxation.** The higher the score, the higher the tax rate. A 1 indicates no income taxes or a flat tax on income of 10% or less, and limited or no taxes imposed on corporate profits.
- **Government Intervention in the Economy.** The higher the score, the greater the intervention of government in the economy. A 1 indicates low government consumption of the nation's GDP—10% or less—and virtually no government-owned enterprises.
- **Monetary Policy.** The average inflation rate over time is the report card of government monetary policy. Inflation of 6% or less is rated at 1; inflation over 30% rates a 5.
- **Capital Flows and Foreign Investment.** This classifies the barriers to foreign investment from very low (1) to very high (5), factoring in the foreign investment code, foreign ownership restrictions, and the flow of money in and out of the country.
- **Banking.** A companion rating to the Capital Flows and Foreign Investment listed above. Very low restrictions are indicated by a 1. A 5 means that the financial institutions are either chaotic, operating on a primitive basis, government-controlled, or corrupt.

- **Wage and Price Controls.** Wages and prices determined by supply, demand, and other market forces rate a 1; wages and prices completely controlled by the government rate a 5.
- **Property Rights.** Private property guaranteed by the government, with an efficient court system that enforces the law, rates a 1; nonexistent property protection or the outlawing of private property rates a 5.
- **Regulation.** A 1 indicates that regulations are straightforward and applied uniformly to all businesses. Regulations are not a burden; corruption is nearly nonexistent. A 5 indicates government discouraging the creation of new business, rampant corruption, and randomly applied regulations.
- **Black Market.** Those rated 1 have a well-functioning free market economy, with a black market at a low level in such things as drugs and weapons. A 5 indicates a black-market economy at a high level, often bigger than the formal economy.

Subcolumn 1 provides the composite ranking for these ten indicators; subcolumn 2, the individual ranking for trade policy; subcolumn 3 indicates foreign investment considerations.

A summary of each table follows.

The Large Seven

The combined annual average population and consumption growth for the seven large nations shown in Table 4.1 is 5.3%. Japan and the United States, large mature markets, are on the low end at 2.0 and 2.5% respectively. China is on the high end with 10.7%; the consumption rate is 9.7 of that total. The Index of Economic Freedom averages 3.0 overall, with the United States at 1.9 and Japan at 2.1. China ranks highest at 3.8, followed closely by India and the Russian Federation in that order. Also note that the trade policy is a 5.0 for China and India, with Brazil and the Russian Federation at 4.0. The large seven have well-supplied markets, or significant barriers to entry, or both. While opportunities may exist, they are in extremely competitive situations (United States and Japan) or carry a high degree of risk (China, India, Brazil, and the Russian Federation). The factors concerning Indonesia are unknown.

The Three Midsize Nations

For the nations shown in Table 4.2, per annum population and consumption growth averages 1.3% and 5.4%, respectively, for a combined total of 6.7%. This is considerably higher than the combined 5.3% of the large seven. The Index of Economic Freedom averages a 3.0. However, Myanmar (formerly known as Burma) is currently waging a civil war, and its trade policy rating of 5 makes it off-limits for even the most hardy gambler. The Index of Economic Freedom rankings for South Korea and Thailand make trade and manufacturing appealing, provided the economic problems present at the closing decade of the 20th century are resolved.

Table 4.1: Population, Consumption, and Economic Freedom

The Large Seven

Country	1996 population (thousands)	Population growth ratio % per year 1990–1995	Private consumption per capita av. annual growth rate 1990–1995	Index of Economic Freedom		
				Overall	Trade policy	Foreign investment
Brazil	161,365	1.4	2.9	3.4	4.0	3.0
China	1,215,414	1.1	9.6	3.8	5.0	3.0
India	945,121	1.8	2.8	3.7	5.0	3.0
Indonesia	197,055	1.7	5.5	2.9	2.0	2.0
Japan	125,761	0.3	1.7	2.1	2.0	3.0
Russia Fed.	147,739	0.1	5.9	3.5	4.0	3.0
United States	265,284	1.0	1.5	1.9	2.0	2.0
Unweighted Average		1.0	4.3	3.0	3.4	2

Source: 1998 World Bank Atlas, pages 24–25; 1998 Index of Economic Freedom.

The Fifteen Small Nations

In the nations depicted in Table 4.3, population growth and growth in consumption rates are outstripping those of the large and midsize nations described in Tables 4.1 and 4.2. The composite ranking of 7.4% is higher, and six of the nations rank substantially higher than that. Uruguay and Chile lead the group, with 10.0 and 8.7%, respectively. Seychelles has a composite growth rate of 15.0%, but no data for the Index of Economic Freedom is available. Chile, Ireland, and Singapore outstrip both the U.S. and Japan in relative composite growth and are competitive when measured relative to the Index of Economic Freedom.

What does all this mean? Why have some nations such as New Zealand, Canada, and the nations of Western Europe been excluded? Does this mean that opportunities are limited in these nations?

The absence of some nations from the tables, such as New Zealand and Canada, indicates that the nation's domestic markets are oversupplied. It also means that export opportunities may be limited by the ability of the nation to compete for markets with competitors located within a market or who have a competitive freight or currency advantage.

Table 4.2: Population, Consumption, and Economic Freedom

Three Mid-Size Countries

Country	1996 population (thousands)	Population growth ratio % per year 1990–1995	Private consumption per capita av. annual growth rate 1990–1995	Index of Economic Freedom		
				Overall	Trade policy	Foreign investment
S. Korea	45,545	1.0	6.1	2.3	3.0	2.0
Myanmar	45,883	1.7	3.4	4.3	5.0	4.0
Thailand	60,003	1.3	6.6	2.4	3.0	2.0
Unweighted Average		1.3	5.4	3.0	3.7	2.7

Source: 1998 World Bank Atlas, pages 24–25; 1998 Index of Economic Freedom.

The criteria for inclusion in the three tables include growing demand and an Index of Economic Freedom that is favorable or has potential for improvement. A business opportunity can emerge as a trading transaction or an in-country manufacturing opportunity, or both. Poland and El Salvador are examples of nations shedding the dictatorships and managed economies of the past and turning to meet the demands of their people. Kronopol's new OSB plant in Poland is a remarkable example.

This Zary, Poland, site has ready access to one of the largest forest resources in the region. Kronopol exports OSB to 18 countries, including Japan. The OSB capacity, combined with the particleboard and MDF lines, will result in a massive plant site production capacity of one million cubic meters. The economy of scale, the access to adequate timber supplies, the operating costs of the facilities, and the ready access to port facilities make this Kronopol facility a world-class competitor.

Ireland is another, as acknowledged in this press notice:

> Much is being made of the Irish economy these days and rightly so. Construction is booming and builders on the banks of the Liffery—and throughout Dublin for that matter—are rubbing their hands with glee. Timber processors and suppliers are also uncharacteristically cheery, in Ireland at least, reaping the benefits of the construction industry's great good fortune. ("LP claims Irish . . ." 1998)

Earlier in the decade, Louisiana Pacific built an OSB plant in Waterford, Ireland. After several years of losses, the plant has become profitable, thanks to the robust Irish market and offshore marketing of excess production. The plant is now reaching onto the continent in European distribution.

Table 4.3: Population, Consumption, and Economic Freedom

Fifteen Small Nations

Country	1996 population (thousands)	Population growth ratio % per year 1990–1995	Private consumption per capita av. annual growth rate 1990–1995	Index of Economic Freedom		
				Overall	Trade policy	Foreign investment
Chile	14,419	1.6	7.1	2.2	2.0	2.0
El Salvador	5,810	2.4	4.7	2.5	3.0	2.0
Ireland	3,626	0.6	3.4	2.0	2.0	2.0
Israel	5,692	3.3	4.2	2.8	2.0	1.0
Jamaica	2,547	1.0	4.6	2.6	2.0	2.0
Lebanon	4,079	1.9	6.6	3.3	5.0	3.0
Maylasia	20,565	2.3	4.8	2.4	5.0	3.0
Mauritius	1,134	1.2	3.7	3.8	5.0	3.0
Nepal	22,037	2.7	6.1	3.4	3.0	4.0
Poland	38,618	0.2	4.6	2.9	2.0	2.0
Saychelles	77	1.5	13.5	NA	NA	NA
Singapore	3,044	2.0	5.1	1.3	1.0	1.0
Slovenia	1,991	–0.1	5.5	3.1	4.0	2.0
Sri Lanka	18,300	1.2	4.7	2.5	3.0	3.0
Uruguay	3,203	0.6	9.4	2.7	2.0	2.0
Unweighted Average		1.5	5.9	2.7	2.7	2.4

Source: 1998 World Bank Atlas, pages 24–25; 1998 Index of Economic Freedom.

Opportunities are there for the finding, and not just in Ireland and Poland. Tariff barriers are falling, and the variety of new and old raw material types gives the manufacturer an opportunity to meet the local market with enough left over to compete in their region. No longer will a patch of forest be the dominating competitive factor. Those businesses that can manufacture close to the consumer, and export excess production to others, will provide the success stories of the new century.

5 The Forest: In Pursuit of Value and Sustainability

A forest represents many things to many people. It is much more than a wooded tract of land. The ongoing dialogue about forests and forestry has transcended its rural domain and become a topic of concern for an astonishing number of people in an equally astonishing number of forums, raising numerous questions: Just what is a forest? What benefits should be derived from the forest? Who should receive these benefits? In what quantity? Over what time frame? What are the rights and expectations of the landowner? Of the community at large? Of the nation? Of the world? Who should make the allocation decisions? How does the landowner pursue value and sustainability?

The answers can be reached by understanding the past and its implications for the future, interpreting the current operating environment, taking advantage of space-age developments that aid the practice of forestry, planning for the future, and then working the plan. The steps are same, whether managing a small plantation or a huge forest.

The Past and Its Implications

In 1900, at the dawn of a new century, the horseless carriage was a novelty, and kerosene lamps were still in vogue. The United States had a largely rural population of about 38.8 million males and 37.2 million females, and the greatest use of wood was for fuel. Times have changed; the frontier era of settlement in America has ended, and the use of wood as fuel has nearly come to an end. However, the nation's appetite for wood-based products is greater than at any other time in our history.

Currently the United States produces about one-quarter of the world's wood fiber but uses approximately one-third of all wood consumed around the globe each year. According to the Food and Agriculture Organization of the United States (FAO) the annual U.S. per capita consumption of wood is about 2.5 cubic meters per person per year—more than three times the global average of 0.7 cubic meters. Developing countries use even less—only about 0.5 cubic meters per person per year, most of it used as fuel to cook food and keep warm. Fuelwood demand, as recently as 1995, consumed about 54% of the global wood supply. Unfortunately, fuelwood still comes primarily from the natural forest, or lands that were once natural forests.

Dr. James McNutt, an international forestry consultant, projects a fiber shortfall in Pacific Asia by the year 2005. A Pacific Asia annual deficit of about 175 million cubic meters will cause an overall Pacific Rim shortfall. The total regional deficit is expected to be about 65 million cubic meters per year. This figure includes the forecast yield of the North American and Siberian forests.

The Pacific Rim contains a large portion of the world population; it also contains some of the world's most productive forestland. But it is only a portion of the total, albeit a significant portion. With the demand for wood fiber increasing worldwide, forest and forest management issues are becoming urgent topics. The need for wood fiber, plus the need for other benefits of the forest—to the air we breathe, the water we drink, the biodiversity we demand, the indigenous culture preservation we seek, and the vistas we enjoy—will require understandings of epic proportions. The search for common ground on how we think about the forest and how we proceed will dominate discussions and negotiations in forums that transcend regional and national boundaries.

Looking Forward: The Lessons of the Past

In order to find that common basis for moving forward, we must understand the forestry lessons of the past, how they relate to the present, and how we envision the future. What then are the lessons of the past, and how can all of us benefit from these lessons? This section explores the Ten Lessons of the Past depicted in Exhibit 5.1.

Change Is Constant in the Forest

Some years ago, the author was a guest speaker before a group of U.S. Forest Service planners. The previous speaker had described the ideal forest as one exactly like the

Exhibit 5.1: Ten Lessons of the Past

1. Change Is Constant in the Forest
2. Forests Are a Limited Resource
3. Trees and Forests Are Renewable
4. Demand Is Growing
5. Different Forests and Needs Require Different Solutions
6. Forest Decisions Are Political Decisions
7. Public Opinion and Forest Management May Conflict
8. Solutions Are Best Built from Common Ground
9. There Are No Secrets in the Forest
10. Innovation and High Technology Are Essential

North American forest at the time of Christopher Columbus. To her, this was the ultimate state of the forest in western Oregon, and the United States generally. She failed to recognize that most of the western Oregon forestlands had been burnt over one or more times in the centuries before man's arrival, and that insects, disease, and weather disturbances have periodically changed the face of the forest. Yes, there were patches of timber that were centuries old—but these were only parts of the whole.

My colleague's perception of the concept of a balance of nature or a natural equilibrium has long held sway in academic and ecological circles. The idea has such emotional appeal that it remains a barrier between those that believe versus those who don't.

However, the accumulation of scientific evidence has led many ecologists to abandon the concept or declare it irrelevant. Dr. Steward T. A. Pickett, a renowned plant ecologist at the Institute of Ecosystem Studies of the New York Botanical Garden at Millbrook, N.Y., said " . . . [it] makes nice poetry, but it's not such great science." (Stevens 1990)

Dr. Simon A. Levin, a Cornell University ecologist and president of the Ecological Society, expressed the changing view:

> "There will always be people who will cling to old ideas, but certainly the center of mass of thinking" among ecologists has shifted away from equilibrium and "toward the fluctuating nature of natural systems." (Stevens 1990)

Mankind is in a constantly evolving state. Forests, like mankind, have their natural life cycles and thus should be respected. The most compelling issue facing us at the eclipse of this century is the need to achieve a unique balance between utilizing the forest as a helpful resource for man and allowing the forest to play out its natural life cycle. This may entail educating those who do not understand nature's complexity.

Forests Are a Limited Resource

Deforestation is occurring in the majority of nations, according to the 1998 World Bank Atlas. While detailed and exact information has not been prepared, it is clear that the world's timber base continues to shrink.

> . . . much of the world's biological diversity is in developing nations, and it is estimated to be disappearing at 50 to 100 times the natural rate. Wetlands and forests are being lost at 0.3 to 1.0% a year. (*World Bank Atlas* 1998, 26)

The 1.0% figure is an estimated average, taken from data for the years 1990 through 1995. Many nations are deforesting the land at measurable higher rates, particularly the smaller developing nations. However, there are some notable turnarounds. China is now in balance and appears to be gaining in the reforestation effort. The United Kingdom, France, Ireland, and the United States are other notable examples of nations achieving gains in forested areas. The task ahead for each nation and the world community is to balance growth and harvest, and move ahead to place idle or underproducing lands back into use as natural or plantation forests.

Drab scrub-brush savannah lands being replaced by plantation forests.

Trees and Forests Are Renewable

Today, people can do more than reforest the land; we can create a forest on sites not traditionally used for tree growing, and harvest them in little more than a decade. As harvests on traditional tree-growing lands are being reduced or curtailed, the new forests in Brazil, China, Argentina, New Zealand, Australia, the United States, and other countries provide a growing volume of wood fiber and other benefits. The Brazilian forest illustrates this effort:

> The drab scrub-brush savannah lands of the Brazilian forest seemed to go on forever as the light plane left the Amazon River behind and flew toward French Guiana. Suddenly, rectangles of green appeared within the savannah lands below. The Amcel forester explained that it was a new forest, little more than a decade old, planted with Caribbean pine (*Pinus caribaea*) and a species of eucalyptus. The forest, managed using a compatible mix of species for the soil and the region, uses an ongoing genetic upgrading of the selected species, and advanced silvicultural techniques. A clear-cut harvest was in operation, with the logs trucked overland for processing at a chip mill. The pulp chips were a direct replacement for wood fiber produced on more environmentally sensitive sites. Papermakers in Europe were the key customers. (Author's field notes)

Some observers continue to be critical of the so-called man-made forests. They overlook the fact that the benefits of these forests outweigh the perceived disadvantages. Each new forest reduces the pressure on natural forests and represents a significant contribution to the wood needs of the world. The earth has a remarkable capability to restore, change, and adapt to the intervention of man.

Managing a variety of forest types, for an overall balance of forest values, is our emerging responsibility. The man-made forest contributes to a real and growing need for wood fiber. Although plantation forests currently account for only about 3% of the global forest area, this forest type represents a major opportunity to increase industrial wood supply.

Demand for Wood and Other Forest Benefits Is Growing

Population growth, consumption growth, and the real and perceived prices for wood fiber and other forest benefits are the primary drivers of consumption. The heightened demands of the populace also affect the forest supply base through agricultural and other nonwood needs and wants. Can we meet the demand for wood products with our ability to find fiber?

It will take something more than growing trees faster on an expanded land base. It will necessarily include recycling, use of alternate processes and materials, and wiser use of the raw fiber resource. Discovering an alternative energy source for the wood traditionally used as a fuel, for example, will be essential to effective forest management.

Differing Forests, Differing Needs Require Different Solutions

The controversy over whether or not to harvest continues to rage. Huge tracts of lands, even whole forests, have been declared off limits, as the "no harvest" advocates still dominate the agenda in the developed nations. Forest management cannot be based on an economic or noneconomic model alone; single use is not an option for most forests. We must strike a balance between the uses and benefits.

Jean Mater, in her book, *Reinventing the Forest Industry,* came to this conclusion:

> Activists will not condone producing wood for shelter as justification of clear-cutting and other forest industry practices they dislike.... One consequence of a politically involved society is that technological advances are not accepted in lieu of social responsibility or a social ethic. (Mater 1997)]

The language of forest management grows more value-laden as the interested parties try to describe their hoped-for outcomes. Silvaculture—a forestry term used to describe the theory and practice of forest establishment, composition, and growth—is being replaced in some quarters by a new term, ecoculture. Encompassing a much wider range of goods and services, ecoculture addresses ways to " . . . manipulate the entire forest ecosystem to diversify, finance, organize, and sustain its aggregate productive capacities." (Case Studies, pp. 1–4)

Coping with a new language, differing needs, and changed expectations will test even the most resilient forest professional.

Forest Decisions Are Political Decisions

Government- and citizen-mandated forest practices, standards, and regulations, land tenure and harvesting policies, environmental laws, and tax legislation and policy

There is a growing move away from clear-cutting.

all shape forest management decisions. Whether based on coercion or incentives, government policies directly affect the composition and growth of forests, and harvesting practices used there.

> Citizens are increasingly important in forestry. Not only do they participate in public consultations in specific cases, but they are also becoming a part of daily decision making and the planning of forestry operations. Clear-cutting, monocultures, prescribed burning, roads, pesticides, and roadside debris . . . in all these issues the public has the power to shape practices. (Guimier 1998)

In 1996 the voters of Maine were given three choices during a general election: ban all clear-cutting, allow nongovernmental organizations (NGOs) to be watchdogs of forestry practices, or reject both options. Fortunately for the forest industry, the citizens rejected both alternatives. The option requiring NGO supervision could appear on the ballot again; the outcome could result in NGOs gaining equal standing with taxpaying landowners in deciding forest policy and practices in Maine. Two years later Oregon's Measure 64, an initiative banning a variety of forest practices, including clear-cutting, became a prime topic of concern for the forest landowner in that state. It too was defeated, but is expected to resurface at a later election.

Jim Melican, Executive Vice President and General Counsel for International Paper, summed up the landowner's situation:

> Events in Maine are telling us it is essential that we sit down at the table with people we once considered adversaries and start building partnerships for the

> future. I'm talking about partnering with elected officials, with regulators and with mainstream environmentalists.
>
> I am not preaching brotherhood; I'm preaching survival. If we fail to redefine ourselves, we may very well be legislated and regulated into a noncompetitive position on the world scene. (International Paper 1997)

The landowner and manager must deal with the growing expectations of citizens and consumers who are a part of the global effort to regulate forest practices. Environmental issues, such as global warming and the effect of deforestation on the planet, are the focus of a number of multinational conferences sponsored under the auspices of the United Nations. The idea that the actions of one nation or its constituents affect the livability of the whole planet is gaining rapid acceptance in the global community. Chapter 17 discusses the politics of forestry in more detail.

Public Opinion and Long-Term Forest Management Decisions Often Conflict

There is an inherent conflict between transitory public opinion and long-term forest management. Public opinion reacts to current events, whereas forest management decisions are based on a growing cycle of decades. This conflict has existed for over a century, as evidenced by this 19th-century legislative initiative:

> The New York Constitutional Convention of 1894 proposed an amendment to prevent any timber on the State Forest Preserve from ever being sold, removed, or destroyed. That amendment was born of well-founded distrust of the politicians in Albany and the Commission in charge. It vetoed selling out to the lumbermen (which had its advantages), but it vetoed Forestry as well. Nevertheless the people approved it overwhelmingly. (Pinchot 1998)

The current call for a harvest moratorium on federal lands in the United States also vetoes forestry. It is another expression of transitory public opinion. The notion that the problems of the past can be resolved through suspending all harvesting is ill-conceived. Unfortunately, public opinion is like a pendulum; it must complete its swing before coming back to center.

Solutions Are Best Built from Common Ground

Solutions are best built from common ground, not from the power of confrontational advocacy. Oregon Senator Mark Hatfield expressed this well:

> There is no room for simply upping the ante and playing forest politics the way it has always been played. There is room for environmentalists to walk a mile in the unemployed logger's boots, to feel the fear and anger that comes from having your home, your means of making a living—literally a way of life—ripped out from under you. There is room for the logger to realize that many of the ecological challenges we face are very real and that the concern for those challenges

Old-growth Alaskan forest being renewed.

expressed by the environmentalist is sincere, and there is room for both sides to recognize how inextricably we all are tied to all of Oregon's wondrous resources. Like it or not, we are all inhabitants of the same planet. (Baldwin 1993)

The senator refers to Oregon, but his comments apply to the world at large. The approach to finding a common ground was suggested by Gifford Pinchot a century earlier:

> There were two possible ways of going at it. One was to urge, beg, and implore . . . to stop forest destruction and practice Forestry; and denounce them if they didn't. This method got onto the platform and into the papers, but it never got into the woods. It had been followed for at least a quarter of a century, and still there was not a single case of systematic forest management in America to show for it. The other plan was to put Forestry into actual practice in the woods, prove that it could be done by doing it. Prove that it was practicable by making it work. (Pinchot 1998)

Both Hatfield and Pinchot offer solutions for finding common ground. Hatfield's "walking a mile in your adversary's boots" sets the stage for Pinchot's older solution, "walk your talk." Unfortunately, today participants spend too much time and money talking. History tells us we would be well served by doing more walking

There Are No Secrets in the Forest

There are no secrets in the forest; the specifics of the changes man and nature have wrought are transparent to the careful observer. A regional governor of a Southeast Asian country learned that when seeking outside assistance in building up the local timber processing industry. A prospective investor asked, "What is the sustainable yield? Are there any environmental or other development issues that could impair the timber base?"

The regional governor and his forestry people assured him they could sustain an annual minimum timber harvest of 50 thousand cubic meters to a high of 80 thousand cubic meters. The investor tried to make his own appraisal, but the lack of roads, the huge area, and the difficult terrain made the task difficult, even with the use of a helicopter. Maps and field data were of some help, identifying a tiger preserve and a reservoir site. Financing might be secured from the nation's World Bank development fund, but only with guaranteed protection of the wildlife preserve and a sustainable harvest of the forest.

Once the prospective investor was able to purchase recent satellite photos, he was able to make a rapid, cost-efficient assessment of the timbered area. Compared to recent maps and field data, the photos indicated that the timber base was far less than expected. No more than 10 to12 thousand cubic meters could be harvested sustainably, The existing maps had significant errors. And finally, the photos showed that a logging contractor had recently harvested within the boundaries of the wildlife preserve.

Some governments and forest managers still believe they can keep forest secrets from the public. However, the technical capability to audit the forest and its changes are becoming less expensive and more sophisticated. Enlightened land managers around the globe must accept that the forest's true condition is transparent to all those who want to know.

Innovation and High Technology Are Essential

The forester now has access to an array of high-tech tools—among them global positioning systems (GPS), electronic data recorders, and electronic distance measurement devices. The Forest Engineering Research Institute of Canada (FERIC) is currently developing a computerized model called Interface that allows managers to simulate various harvesting and regeneration scenarios and calculate their total cost. Computer graphics picturing a tree before it is harvested, with optimization similar to that found on an edger optimizer in a sawmill, is now available to the cutter. These and other decision-support tools will control operations within environmental, ecological, social, wildlife, and other parameters that aid the harvesting of wood.

Forestry managers will still make the final decisions, but these high-tech tools simplify the task of accurate decision-making, taking much of the guesswork out of forestry.

The Road Ahead

The Southern Hemisphere could not have emerged as a major source of conifers without a marriage of innovation and technology. Inventive managers developed this new forestry sector through a combination of silvicultural practices of planting and pruning, genetic improvements, and advanced harvesting practices. In the Northern Hemisphere such innovative efforts run into two major barriers: the enduring traditions of forestry and the more recent thinking about the practice of forestry. Traditions include beliefs that (1) longer rather than shorter rotations are most desirable, (2) use of fewer rather than more forest entries is the way to grow the forest, and (3) an undisturbed forest is the best forest. It's thought that every forested acre should provide a full range of benefits, often with wood production being just one of the many. Then the talking starts—in the media, in the courtroom, and in whatever forum we can find—because the old (and sometimes the new) approaches don't work very well

Forestry and wood products are long-term businesses; the participants have no choice but to plan for future laws, regulations, and incentives. As forest users' expectations change and the demand for products increases, no one can say exactly how forestry in the new millennium will differ from the past. But the following is certain:

- The public will demand more say in determining how a forest is managed and how its benefits are prioritized.
- The decisions made will have a profound effect on local, national and global affairs.
- Along with a shift away from wood as a fuel, use of wood in more engineered applications and in combination with nonwood substances will increase.
- It will be necessary to obtain more wood products from an ever smaller land base.
- Plantations will be an increasingly important wood supply source.
- Demand for wood and other forest benefits will grow ever faster.

The road ahead will be exciting for some, frustrating and painful for others. The successful forest industry participant will be adaptive, innovative, and resourceful. These qualities are prerequisites to business survival.

Genetic Engineering and the Forest: The Application of Space-Age Science

6

In the coming decade, genetic engineering will rival computers and engineered wood products in importance to the worldwide forest industry. Genetic engineering is a hot, controversial, and little-understood topic. It carries a host of legal, environmental, ethical, and other problems. But the forest industry needs what genetic engineering can do for it, and sooner rather than later.

Teams of foresters and biotechnology scientists are developing genetic engineering techniques that enter a tree's cells, extract genetic code for its sought-after features and structural characteristics, and then lock these characteristics into the growth of a new forest.

The forest industry participant faces many tasks: first, gaining an understanding of this new and exciting technology: second, recognizing the benefits of growing better trees in a shorter time frame: third, pushing forward the frontiers of genetically based forest science by supporting the efforts of the scientist and forester teams: and fourth, incorporating emerging technologies on available tree-growing lands.

Trees and Genetics

Both a tree and a forester are built of and function by means of molecules, and the two have still more in common. Molecules, amazing building blocks, are nothing more than combinations of a half-dozen elements: carbon, hydrogen, nitrogen, oxygen, phosphorus, and sulfur. Every living thing is constructed with molecules, and at the molecular level each living thing functions fundamentally in the same way, be it a forester, a tree, a goldfish, or an earthworm.

The story doesn't end there. The structure of genes and DNA is elegantly simple, yet hopelessly complex for the nonscientist. What is important is this: a defined structure determines the species and its attributes. This structure can be redefined and a new "blueprint" created, and these combinations are limited only by our knowledge, our thought processes, our desire to model and test, and the limitations of nature itself.

> As a result of scientific exploration we have discovered that inside each living organism there is a "recipe" with endless combinations that makes every

creation of plant, animal and human different from every other. It is imprinted on a substance in the cells known as DNA (deoxyribonucleic acid). This DNA has small particles called genes which carry the instruction codes for the characteristics of all life forms. A single plant has about 100,000 genes which direct development through its life cycle. (*Agbiotech Infosource,* Issue 8)

Scientists have found ways to identify genes with special characteristics and transfer them from one type of life form to another. They have begun redesigning the attributes and characteristics of trees and other life forms.

Engineering a Tree

Genetic engineering, and mass duplication of the results, can produce improved strains in a matter of months that once took years to develop. Using a rapidly expanding knowledge base while identifying the right fingerprints, tree biotechnology breakthroughs have opened up endless ways to improve species and species attributes. Two scientific procedures have opened the door to improvement: identifying the tree's genetic characteristics, and designing new characteristics.

Identifying Genetic Characteristics

The characteristics, or fingerprints, that determine a species' genetic makeup are the same for every tree cell within the tree being studied. These fingerprints look much like the bar codes used by grocery store scanners.

Scientists follow these steps to obtain a fingerprint:

- **Gather a DNA sample.** Small samples—usually small leaf or needle segments—are taken.
- **Treat the samples.** The samples are treated with chemicals to extract DNA from the cells.
- **Add enzymes.** Enzymes are proteins that promote chemical reactions; added to the DNA, they function like scissors to cut the DNA into millions of fragments of various lengths.
- **Place fragments on a bed of gel and apply electric current.** The current sorts the fragments by length and organizes them into a pattern.
- **Transfer to a nylon sheet.** By placing the gel against the nylon sheet, the DNA pattern is transferred.
- **Introduce a probe, or short strand, of radioactive DNA to the pattern on the nylon sheet.** The probe binds to specific DNA fragments.
- **Expose the nylon sheet containing the radioactive probe to x-ray film.** Dark bands, resembling a bar code pattern, emerge at the probe sites.

The DNA of each individual tree is unique, producing a unique set of fragments. The resulting fingerprint can be used to identify the specific tree (whether or not this progeny is carrying the traits of its parents) or, in some cases, identify disease or resistance to disease. This fingerprint serves as a benchmark—the starting point for a range of uses limited only by the imagination of the researcher.

Cooking Up New Plant Characteristics

Identifying the fingerprint of the tree is the first step. This knowledge guides the "cooking up" of new tree attributes, attributes intended to benefit both man and nature. The "cooking up" steps are similar to the fingerprinting process, at least at the start.

- **Obtain a DNA sample.** Technicians use leaves and needles wherever possible, since they have the most living cells. The next best choice is the cambium layer. The sapwood has few living cells, the heartwood even fewer.
- **Refrigerate, add buffers, and make a "smoothie."** The resulting concoction is prepared for the next step.
- **Isolate DNA and its counterpart RNA through differential extraction.**
- **Place the genetic components in test tubes: DNA in one, RNA in another.**
- **Process the DNA.** Just as in fingerprinting, technicians use enzymes as microscissors to cut the gene strands into convenient sizes.
- **Create a genetic library.** Technicians isolate tree DNA from all other DNA, and isolate the "interesting" tree genes.

This is an oversimplification of a complex process, but it provides a glimpse of the methods used to identify and isolate genes. The processes and techniques outlined here, and a host of others, provide the tools to re-engineer a tree. Genetic engineering techniques allow us to rearrange the tree's genetic structure, to exactly replicate, or "clone," a tree, and even to transfer a gene from one organism to another. The process can literally cook up a new tree.

Re-engineering a Tree: The Results

Some of the media stories about genetic engineering sound almost like science fiction, rather than fact. The new engineered trees grow so quickly, they might come from another planet. Gmelina (*Gmelina arborea*), a tropical hardwood, is grown in the Gulfito region of Costa Rica. When grown on optimal sites, these trees may have a diameter at breast height (DBH) of about 9 inches (23 cm) at 2½ years. Hybrid cottonwood trees may reach 70 feet (21.34 meters) in 6 years, with a trunk diameter described as "about the size of a soccer ball." One genetically engineered eucalyptus tree species, grown in Indonesia, springs to 38 feet (11.58 meters) in just 12 months—three times faster than the nonengineered parent species.

CHARACTERISTICS OF RE-ENGINEERED TREES

Resistant to pests, infectious diseases, and drought

Shorter rotations

Improved wood quality

Exhibit 6.1: Characteristics of Re-engineered Trees

These trees—the gmelina, the cottonwood, and the eucalyptus—have been re-engineered to grow rapidly. They have also acquired a variety of other attributes, such as resistance to pests, infectious diseases, and drought. The longer rotations of the northern softwoods—the major global wood source for generations—are now being supplemented with the decade-or-less cycles of the new "forests."

For example, Boise Cascade has planted 10.2 million hybrid cottonwood trees on 17,000 acres (6900 hectares) in northeastern Oregon once used as potato fields. The yield from this high-tech forest is an alternative to the conventional softwood sources in the region.

Boise Cascade and others are trying to inoculate their lands from the no-logging agenda of others. They emphasize that the trees are being harvested, not logged. Mechanized harvesting demonstrates the point: a feller-buncher extends hydraulically activated arms around the tree trunk, and severs the tree at the base. The feller-buncher then pivots with the tree stems, and stacks neat piles of harvested trees. A front-end loader then transports the trees to a nearby chipper.

The most aggressive tree re-engineering projects usually result in the fiber being converted initially into chips. However, once the short-rotation plantation is established, it doesn't take long for the wood products manager to identify the new wood source as suitable for lumber or veneer. Changing from fiber to a sawlog rotation usually means adding a few years but very little cost. Often, the same tree species is genetically improved to create a higher wood density, a self-pruning bole, or a larger diameter with less height.

The uniquely Americanized paulownia (*Paulownia elongata* Carolinia) is a hardwood subspecies just now gaining attention in the southeastern United States. It was once a relatively slow-growing tree with poor form; now some call it a supertree. It is versatile; it can be peeled, sawed, or chipped. Light in color, light in weight, paulownia is dimensionally stable; it makes excellent furniture, decorative moldings, and veneer. Since being genetically improved, it is reputed to average 300 board feet per stem when reaching maturity at 10 years. Harvests of 85.5 MBM (Doyle log rule basis) per acre have been reported.

How about the softwoods? Genetic engineering has also brought about shorter rotations and improved wood quality in this group, but at a slower pace. For example, over the last quarter-century genetic improvement of loblolly pine (*Pinus taeda*) has produced shorter growing cycles, better disease resistance, and other improvements. However, the softwoods have a more complicated genome structure (that is, all the genes in a complete set of chromosomes), and results must be tested over comparatively longer rotations. Furthermore, since softwoods are relatively more abundant than hardwoods, researchers have not felt the same pressure to produce results

The genetically re-engineered Brazilian-based Caribbean pine (*Pinus caribaea*) plantations are an example of trends in softwood engineering. Caribbean pine, with 9- and 10-year rotations, is growing on converted savannahs north of the Amazon River. Other fast-growing pines in the south of that nation were introduced to compensate for the disappearance of the natural forest. And then there are the radiata pine (*Pinus radiata*) plantations of New Zealand, Australia, and South Africa.

A tree native to northern California, the radiata pine (better known to Americans as the Monterey pine) was considered unsuitable for commercial use in its native habitat, but breeding and genetic engineering programs have transformed it into a world-class competitor in the international wood markets. The lumber produced from radiata pine has found new value in California as a replacement for the diminishing supply of lumber from the stately, commercially valuable ponderosa and sugar pines.

These success stories are just a few of many; some think the best is yet to come. However, the science of genetics and the application of genetic engineering are not without controversy.

Genetic Engineering: The Issues

Genetic engineering has raised numerous and profound issues for its practitioners, supporters, and critics. They range from emotional issues, such as ethical and environmental concerns, to technical issues of research and application, to very real funding issues, and, finally, commercialization issues. Media attention to ethical and environmental issues frequently overshadow the others.

Ethics and the Environment

The notion of man tinkering with nature is abhorrent to some, regardless of their level of education. The genetic engineering of trees, the emergence of fiber farms rather than forests, the application of an agricultural regimen rather than classical forestry techniques—all are drawing increasing attention.

A genetically improved Caribbean pine (*Pinus caribaea*) stretching to the horizon, Brazil.

There are at least two elements to the controversy. First, questioners express concern that the control of plant characteristics will result in vast tracts of trees that aren't really a forest, just a repetition of a monoculture run rampant. This could happen, but it is not likely. What is happening can be described as follows:

- Genetically improved stock increases land productivity. Further, marginal lands can now grow trees. The ancient forest and other special forested places can remain special, with wood fiber demand met in other ways. The genetically improved tree stands serve as a buffer supply for the growing demand for wood fiber.
- Gene loss through forest fragmentation is prevented. A late-1998 project proposal for Costa Rica typifies what can be done. This study would determine the effect of forest fragmentation on genetic variation. The results will provide baseline data for comparison with future changes in genetic diversity or mating parameters of the same trees and same populations.
- Genetic engineering provides fact-based tools to evaluate the changing use of the forest. In southwest India, for example, rural and urban populations are heavily dependent on a rapidly diminishing forest resource. Genetic engineering can help answer such questions as: What spatial scale of genetic losses may result from deforestation? In what magnitude and degree? What type of forest landscape can best minimize these alterations within a high-use forest?

Genetic engineering and the application of this science can produce facts and solutions that replace ignorance and speculation.

The second concern voiced is that industrial forests will have a massive impact on the environment through excessive use of chemical fertilizers, pesticides, and herbicides. Will these forests be rendered unsustainable, in the long run, if they deplete the soil? The forest practitioners respond as follows:

- Leguminous species are being grown in the tree plantations to improve soil fertility and suppress weeds—an agricultural practice that reduces the need for both fertilizers and herbicides.
- Logging residues and bark are being chipped and left on the land to reduce nutrient loss and act as a mulch, thereby reducing both nutrient losses and water needs, and suppressing weed growth.
- Scientists are identifying genetic improvements that resist disease, insects, and other pests.

Ethical and environmental concerns, coupled with economic pressures, are forcing forest managers to produce more wood fiber, more economically, with less environmental impact—a complex challenge. Biotechnology provides the tools to make this possible.

Technical Issues of Research and Development

Tree scientists and foresters face a number of complex research and development (R&D) issues, including the genome complexity of trees and the need to maintain genetic diversity. Add to this the growing mass of information, and the task of processing, interpreting, and communicating the resulting technical data becomes almost overwhelming.

For example, the research required to identify the genetic patterns—the genome—in a complete set of tree chromosomes is immense. Although the task is difficult with the simpler plant forms, it is even more difficult with trees, and the softwood species' genomes are more complex than those of the hardwoods.

Cheryl Talbert, Weyerhaeuser manager of seed production and timber improvement for western Washington, describes the complexity of the task:

> Conifers have one of the largest genomes (gene patterns) in nature. Foresters who have done much work in this area tend to be pretty humble. For one thing, whatever you come up with, you're stuck with it for 50 years. (Lucas 15)

Mapping the genetic structure of a tree is a huge undertaking. The current state of the research, although vast and growing, is just now making inroads. Some tree characteristics can probably be traced to one or a few genes, but that's the exception rather than the rule. Tree characteristics such as growth, yield, disease resistance, branch formation, density, stem form, and many other traits crucial to tree improvement are determined not by a single gene, nor through a multistep biochemical pathway, but by hundreds and even thousands of genes interacting with each other and with a changing environment. And genome complexity is closely coupled with another complex technical issue: the need for genetic diversity.

Foresters and scientists face a compelling dilemma: What is the proper balance between current economic benefits and the long-term health of the forest; between

Genetically re-engineered Caribbean pine (*Pinus caribaea*) plantation, a close-up.

maintaining genetic diversity over many breeding cycles, and the larger short-term gains possible with the focused selection of only the best genotypes? Intensive selection and breeding programs, assisted by genetic engineering, have the potential to significantly reduce the gene pool—how then can we conserve genetic diversity for adaptive traits?

This question rests in the hands of the geneticist and the forester. There will be no easy answer; the answer may equal the question in complexity.

Throughout North America and the world, dedicated R&D organizations are shouldering the dual burden of doing the research and financing the effort. One such group is formed by geneticists and cooperators in the Pacific Northwest.

The Tree Genetic Engineering Research Cooperative (TGERC) was formed to provide cooperative research on genetic engineering, with this mission:

> The goal of the Tree Genetic Engineering Research Cooperative (TGERC) is to provide research, technology transfer, and education to facilitate use of genetically engineered trees in plantation culture. The TGERC is adaptive, rather than fundamental, in orientation; it seeks to test and develop select innovations in genetic engineering that may have commercial value to wood growing industries. (TGERC Annual Report 1996)

TGERC members include Boise Cascade, Georgia Pacific, International Paper, Weyerhaeuser, and MacMillan Bloedel. A mix of smaller forest products companies, government agencies, academic institutions, and diversified firms such as Shell and Monsanto make up the balance.

TGERC fits a pattern of other organizations with a like mission; research cooperatives usually include forest industry companies, with a mix of academic, government, and other interested parties as members. An academic or government entity usually manages, staffs, and houses the research effort.

The New Brunswick Tree Improvement Council (New Brunswick, Canada) has eight industrial companies, two universities, and the provincial and federal governments as participants and sponsors. Other research institutions include government agencies such as the Institute Of Forest Genetics at Davis and Placerville, California, sponsored by the U.S. Forest Service.

The R&D laboratories mentioned are only a few examples drawn from the vast international research effort. Also worth mentioning are a consortium in New Zealand called "Genes," which focuses on the research effort to genetically improve radiata pine, and the National Forest Tree Breeding Center (Japan), which has a 40-year history of improving sugi (*Cryptomeria japonica*), more commonly called Japanese cedar.

Genetic Engineering and the Computer

Genetic engineering research is just on the brink of studying all the genes in a tree genome—an achievement that will bring us within reach of a complete understanding of how important traits are controlled. Geneticists once wrote project proposals to study just one gene; now they are studying a wide array—often many thousands—at a time. The fields of biotechnology and computer technology are working synergistically, using the information technology of the computer (with its binary code) to unlock the secrets of the DNA sequence (with its thousands of genes).

A plantation of genetically improved trees in Oregon.

The computer does more than process the vast amounts of raw data being generated by the growing number of genome-sequencing projects. The development of sophisticated robotics and computerized miniaturization keeps pace with the complex demands of the research.

The New Millennium: What's Ahead in Tree Genetics?

What's ahead? The easy answer is "you ain't seen nothing yet." The technology is in its adolescence; great discoveries have been made, and they are just the "tip of the iceberg." The industry's slow process of coming to understand biotechnology, and genetic engineering as a biotechnology discipline, is not unlike that of computer users of the late sixties and seventies, trying to understand how a computer functioned and how programming was accomplished. In time the computer became more user-friendly, more convenient, and cheaper. A complex technology became understandable to most.

When it comes to in understanding genetic engineering and its potential, we are still at the point of late-sixties computer users. We have yet to bridge the gap between technology and understanding. But the understanding will come, as remarkable events continue to unfold. What tree-based remarkable events can we expect during the first years of the new century?

- **Timber Supply**
  In the United States, Japan, and other parts of the world, timber inventory is being disconnected from timber supply. Consumers will gradually realize that forest management (or nonmanagement) for scenery and aesthetics, water quality, recreation, wildlife, and other nonwood benefits limits the industry's ability to meet consumer needs.
- **A Dialogue Begins in Earnest**
  Product-producing people seldom talk to tree-growing people, much less the research person. As the timber supply shifts to smaller trees and shorter rotations, and consumer demand continues to grow, an ongoing dialogue between the manufacturer, the forester, and the researcher will become the norm.
- **The Two Heavy Hitters, Plus a Third**
  Engineered Wood Products (EWP) and Engineered Raw Materials (ERM) will be the heavy hitters, with demand continuing to grow exponentially. The recycling of wood fiber will grow as an increasingly important supporting player.
- **Specialized Trees**
  The fruits of genetic engineering will become more visible in the forest, the mill, and the marketplace, with specialized trees tailored for specific processes and products. For example, a tree for the papermaking process may contain decidedly less lignin; trees with less lignin require less time, energy, and chemicals to pulp.

As the worldwide forest industry relies increasingly on engineered raw materials, and on genetic engineering to provide them, both the industry and its constituencies will come to accept these new dimensions in forestry.

Wood Fiber Reused Is the Tree Best Used

7

The harvested tree is still the primary raw material for wood-based-type products, and probably always will be. However, sweeping changes are taking place in the industry, with innovative products and processes gaining acceptance and market share. Recycled wood fiber, agricultural fiber sources, and inorganic substances now supplement the forest in providing raw materials for wood markets. The manufacturer has an expanded materials base, the consumer has a greater choice at a lower cost, and the environment benefits. It's a win-win situation for both the global manufacturer and the consumer. Those industry members who get to know the trends in raw material, processes, and products will better understand supply and market dynamics in the opening decades of the new century.

Beyond Tree Farms and Plantation Forestry

The traditional forest, the tree farms of the earth's temperate zones, the emerging short-rotation plantations of genetically improved species—all of these combined will still fall short of meeting the forecast global demand for paper and wood products. This demand, as well as expected fiber-supply shortages and harvesting restrictions, will prompt sharp price increases. Wood product producers must expand their supply of raw material. But the looming crisis of supply and demand is only one dimension of the problem; environmental considerations also demand attention.

Forests serve a valuable role in preserving the climate. A vigorous, growing forest feeds on the vast quantities of CO_2 produced as a byproduct of our industrial societies. The forest industry faces a dual challenge: to serve the growing demand for paper and wood products while maintaining and enhancing the quality of life on our planet. We must reduce the demand on our forests and fiber farms by recycling wood fiber and by developing earth-friendly alternative supply sources.

Waste Wood Fiber: A Problem and an Opportunity

Today, the developed nations struggle to dispose of an enormous and ever-increasing accumulation of waste products. As landfills reach capacity and shut down, disposal options dwindle in number and increase in cost. Many landfills no longer accept construction and demolition materials, which make up fully half of the total waste generated.

Wood reclamation is nothing new; it has taken place for centuries. As mills process logs into lumber, panels, and other wood products, they generate byproducts in the form of bark, chips, sawdust, and other residues. In early days, mills chipped wood slabs and edgings into pulp chips; later, they used planer shavings in particleboard. The industry has also put waste wood to use as paper, boiler fuel, and animal bedding. But until recently, these only made use of milling byproducts; post-consumer wood fiber leftovers were considered trash. Our society still throws away enormous quantities of wood fiber in the form of newspaper, packaging, and other paper-based waste.

The U.S. Central Intelligence Agency's *Handbook of International Economic Statistics* lists paper production, consumption, and waste paper recovery for 24 nations during the years 1987–1989. These figures, although somewhat dated, indicate progress made—and yet to be made—in waste paper recovery. The five top waste paper recyclers are shown in Exhibit 7.1.

These countries recycled nearly 50% or more of their total paper consumption. In contrast, the U.S. total was 30%, well below the average for all nations surveyed. The timber-producing countries of Canada (21%) and New Zealand (18%) were even further

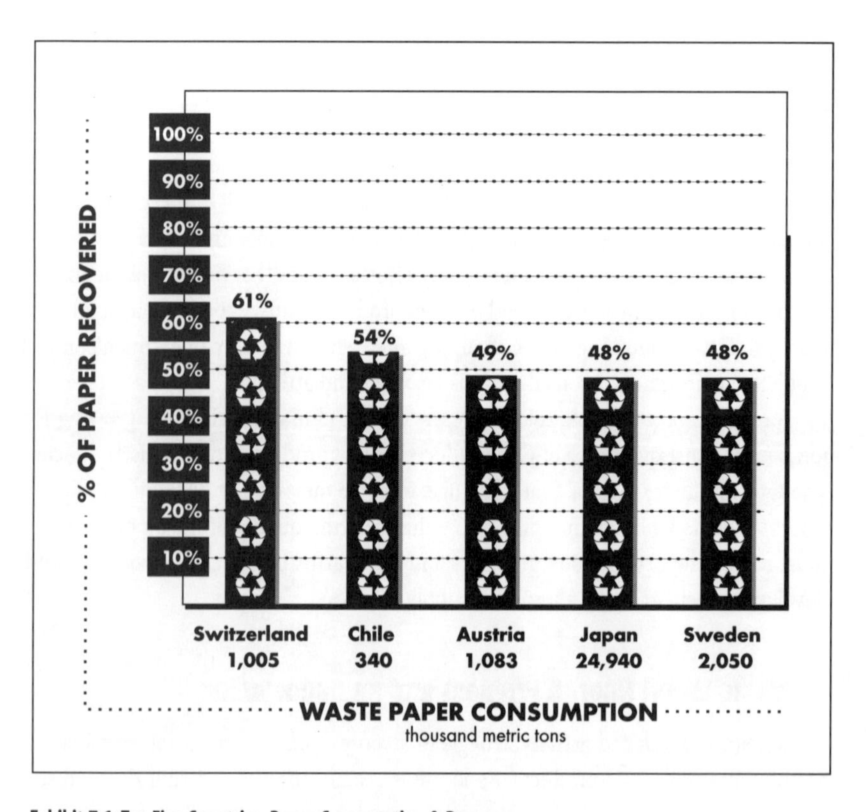

Exhibit 7.1: Top Five Countries: Paper Consumption & Recovery

below the average expected. This exhibit indicates that those countries that see the need to recycle do so.

Much has changed since that survey was made, but much still needs to be done. Wood fiber can be recycled up to five times before the fiber turns to mush in the digester. The most obvious path for waste leads back to the paper mill; however, other uses are developing.

Louisiana Pacific's Nature Guard insulation uses newspapers that are processed into loose insulation. In the typical installation, the consumer or contractor pours the material between the ceiling joists in the attic to a depth that is specified for the R value sought.

Recycled waste paper is a key component of the "urban forest"—a term used to describe the efforts being made to recover and recycle pallets, wood debris, and other like materials near major population centers. A 1998 University of Michigan study, looking into the uses for recycled newsprint fiber, pointed out that manufacturers use the same pulping process to produce thermomechanical newsprint, Medium Density Fiberboard (MDF) fibers, and hardboard fibers. Newsprint fibers lack the coarse screen fraction necessary for the production of high-strength MDF, but they produce respectable properties at high board densities, and show promise for use in highly densified face layers in MDF and particleboard. The Michigan study also notes the ongoing search for additional economic benefits from recycling waste paper.

Wood Reclamation and Recycling

Both the paper and metal industries have been recycling for decades, dating back to the WW II era. Why didn't the wood products industry act sooner to incorporate post-consumer waste wood into new products? First, because timber has been seen until recently as abundant and renewable. Second, the industry is still trying to find economical ways to clean urban waste wood for reuse. But, as shown in Exhibit 7.2, recent trends reflect the growth of wood-recycling opportunities. In 1990 the U.S. municipal solid waste stream included 12.3 million tons of wood, of which an estimated 400,000 tons, or 3.2%, was recovered. By 1993, 13.7 million tons of waste wood was generated—with about 1.3 million tons, or 9.6%, recovered. This dramatic trend shows promise for, the future.

The U.S. Forest Service has taken a leadership role in directing resources to the waste-wood recovery effort, establishing a comprehensive approach that is a model for the U.S. and other countries. The Forest Service identifies opportunities through research, and provides direction for the collecting and processing of waste wood.

A Forest Service publication of March 1994 identified housing and construction applications as offering the greatest opportunities for recycling waste wood.

> In most new homes, very few of the available wood-based products consist of materials recovered from the municipal solid waste stream. . . . Great potential

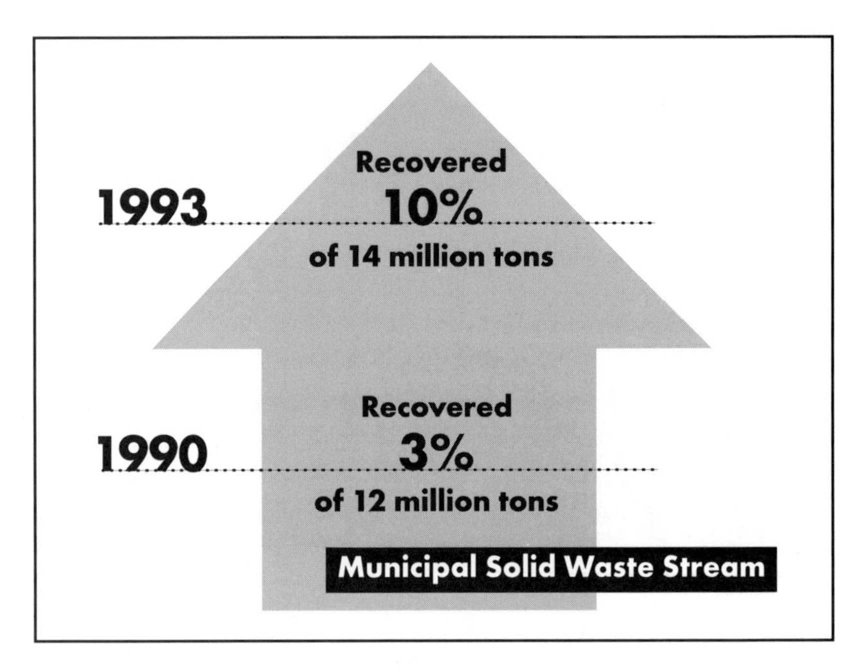

Exhibit 7.2: U.S. Municipal Solid Waste Stream, 1990 and 1993

> exists for providing building products made from recycled waste wood and wastepaper as complements to those currently made from virgin material. (USDA 1994)

The Forest Service aims to replace up to 20% of the virgin wood fiber currently used in housing by the year 2000. The agency plans to allocate resources, develop technologies, and assist private industry to develop products and markets that will consume up to an additional 12 million tons of the paper and waste wood now going to landfills.

Research and development efforts appear to be matching this goal. The July–December 1997 issue of the Madison lab publication *Dividends from Wood Research* lists 22 published studies dealing with wood recycling or disposal. These recycling studies often include other materials, such as plastics, cement, and agricultural products. Most focus on construction products for the home, emphasizing roof systems, foundations, cabinets, and millwork.

The industry hopes to reduce the virgin wood fiber used in roof systems to 40% of present demands. Roof sheathing, trusses, and rafters can be made from recycled wood, supplemented by solid wood structural elements. Recycled paper fiber works well for roofing felt; wood fiber can be mixed with concrete as the basis for roof tiles; and shredded wood mixed with fiberized waste plastic makes a good loose-fill cellulose insulation. These innovations are just a beginning.

Manufacturers of Permanent Wood Foundations (PWP) may soon replace some plywood with an aggregate of wood-cement composite blocks—fabricated using recycled

Urban wood waste provides much of the raw material for this Willamette Industries MDF plant.

treated wood fiber—along with treated composite panels. Waste wood-based composite millwork, such as mouldings and countertops, is already commercially available.

A recent progress report from Oregon State University's Center for Wood Utilization Research (OSU) cited work being done to incorporate wood and plastic into new products. Striving to improve the material properties of recycled wood-plastic composites and expand the choice of such products, the lab is developing a thermosetting, cross-link compatibilization and bonding system. Researchers at the Universidade de Ameiro in Santiago, Portugal, are collaborating with OSU to investigate these new systems.

Waste-Wood Collection

While R&D efforts continue on these and similar projects, the Medium Density Fiberboard (MDF) industry is making giant strides in utilizing waste wood. Willamette Industries is a pioneer in waste wood utilization. Willamette's Eugene, Oregon, MDF plant was one of the first industry plants to achieve industrial-scale production from recycled wood. This modern but relatively small plant, with a 4 × 16-foot press, manufactures MDF in thicknesses ranging from 1/4 inch to 1 inch (6.35 mm to 25.4 mm). It complements the waste wood with some wood material, such as hardwood chips and sawmill/veneer byproducts readily available in the region. The plant gets its daily raw material requirement of 150 to 200 tons from an elaborate processing and transportation system, which obtains wood from sources up to 200 miles away. Such systems are the key to diverting useable wood fiber from the waste flow.

To establish a waste-wood collection network, a producer must first identify the waste generators to determine whether the volume and type can support a factory, or if it would be best used for fuel, floral mulch, and animal bedding. Supply sources include receiving docks where packaging and pallets are frequently discarded, as well as waste from homes, factories, construction or demolition sites, and other sources.

An often overlooked source is the secondary wood industry. Typical secondary manufacturing machine centers that generate waste wood include crosscut operations, double-end tendoners, sanders, and router operations. In-plant waste-wood collection not only reduces the flow into the landfill, it can be a source of byproduct income.

The American Forest & Paper Association, a timber industry trade group, has spearheaded the formation of a waste-wood-processing industry through its publication, *National Wood Recycling Directory.* Published since January 1996 in cooperation with the USDA Forest Service, the directory lists a total of 612 known Wood Residue Receiving Centers (WRRC). Its authors offer this mission statement:

> The biggest obstacle to increased wood recycling is the absence of an infrastructure to salvage and reprocess wood. This is why the *National Wood Recycling Directory* was produced. In providing wood users a listing of where wood residue is being accepted for recycling, the *Directory's* producers hope to increase the rate of wood recovery. (*National Wood Recycling Directory* 1996)

Waste-wood recycling is catching on as demand for the fiber continues to grow. The directory provides an added benefit to users. By publicizing the location of the individual WRRC, it also provides an evaluation of the network and an assessment of what needs to be done to improve wood recovery.

When waste wood is used as a raw material, government regulations generally do not impose specific standards. It is important, however, to check applicable laws and regulations before moving forward to process the waste wood. Most U.S. states have hotlines to help determine which regulations, if any, apply and how to comply with them.

Processing Waste Wood

Setting up a waste-wood process begins with identifying the suppliers, the waste type that's developing, and the products being produced from this waste. Next, the producer needs to design a low-cost process for milling the incoming waste wood into desirable raw material for the product to be manufactured. The incoming material must be presorted; any contaminated waste (with undesirable material that can't be removed cost-effectively) is screened out and processed to hog fuel or other low-value end uses. The sorted, screened remainder flows into a preshredder that reduces the wood to manageable sizes and opens it up to facilitate the magnetic removal of ferrous metal such as nails and bolts. This step yields a material that conveys easily and can be readily handled for transport. Further sorting removes nonferrous wastes such as rock, paper, and plastic. The waste then goes through coarse size reduction, drying, screening, sorting, and further fiber preparation. The processing configuration used is based on the raw material, the size or mix of sizes, degree of contamination, and the products—board furnish preparation differs from preparation for garden mulch.

Currently, board plants use the lion's share of recycled wood fiber. The CanFibre MDF plant at the Port of Amsterdam, The Netherlands, is an example. Typical of the newest plants, it's situated in an urban forest—being adjacent to a large, established population, the plant benefits from ready access to markets and raw materials. It is designed to produce 212,400 M3 (120.0 MM 3/4 basis) annually. The raw material will be 100% waste wood supplied under two separate contracts. Each contract has a 20-year term, with provisions for two 10-year fixed-price periods. The plant is engineered to be an environmental good neighbor: the MDF board has negligible formaldehyde emissions because it uses a PF resin, similar to the resin used in exterior plywood, rather than UF resins that emit much higher levels of formaldehyde.

Harvesting the Urban Forest

Both the Willamette plant and the CanFibre facility described above are harvesting the urban forest. So are the Nature Guard recycled newspaper insulation plants of Louisiana Pacific. Paper mills also use post-consumer waste paper gathered from nearby urban centers. Typically sized for both the resource and the market demand, the mills are located closer to the customer and designed as modern, environmentally friendly facilities, producing a green product. Cheap wood fiber in a variety of forms, low-cost transportation to major markets, modern plants, environmentally appealing products—these factors enable the plants to emerge as major competitors on world markets. And successful trendsetters like these attract other recycled-waste-wood producers to the field.

Wood-based niche businesses are also harvesting the urban forest. The Urban Forest Woodworks of Logan, Utah, offers a variety of logs in different diameters, lengths, species, and wood characteristics. Trees blown down in storms, hazardous or overage trees removed for cause, and trees removed during modernization or expansion all find a home in this wood lot.

Owner George Hassenthaler made these comments:

> I obtain high-figured wood patterns and species that are hard to find anywhere else because it's salvaged urban wood. . . . These species are rare in this part of the country. Many were planted 100 years ago by the pioneers who initially settled the Salt Lake Valley. (*FDM* August 1998)

Hassenthaler estimates that the Salt Lake City landfill receives about 2.0 MMBM of similar logs each year. "It's just getting thrown away. I can't bear to think of that." (*FDM* August 1998)

If 2.0 MMBM is the right number for one urban landfill alone, imagine what the total number is for similar situations in the United States and in older countries in other parts of the world. And logs are just one form of recyclable urban waste wood. The harvesting of urban lumber, sometimes called "ancient timber," is another way. Fine Lumber Company of Brooklyn, New York, salvages timbers from old warehouses, houses, and other structures, then resaws and remanufactures them into new products. Low-grade lumber—often Douglas fir and spruce—is sold for shoring, sewer sheeting, and concrete forming. Fine Lumber uses old-growth longleaf pine, a scarce and treasured species, in furniture, cabinets, and millwork; its beauty and durability is unsurpassed for flooring and decking. Fine Lumber is not a new business— it has been around since at least 1933—and recycling is nothing new to the firm. What is new is the beauty in each timber as the timeworn outer wood is removed, revealing wood characteristics that have become rare as the old-growth forests of North America are either cut out or locked up.

Fine Lumber in Brooklyn, Urban Forest Woodworks in Utah—these companies and others have learned that the solid wood is available; it just needs to be pulled out of the waste-wood stream and diverted into productive uses, filling the need for wood products without touching the natural forest.

The Other Fibers—and the Nonfibers

Innovative producers extend the raw material base by combining a growing variety of other organic fibers and inorganic substances with wood fiber, either prime or recycled. Cement, gypsum, and other inorganics combined with wood fiber in panels meet a variety of needs, mostly nonstructual. Some of these boards have been around for many years; others, such as fiber-gypsum, still seek widespread customer acceptance. The manufacturers are learning that they don't need 100% wood to make a board, and customers are accepting these wood-mineral hybrids.

It is estimated that less than 1% of the world's paper is made from materials other than wood fiber. With the predicted doubling of global paper demand in the next 15 years, paper producers look to wood substitutes with renewed interest. Kenaf, a herbaceous annual native to east-central Africa, thrives in warm, moist areas. The long, woody fibers that make up about a third of its stem have a pulp yield of roughly 60% by weight. A fast-growing crop, kenaf reportedly produces two to three times more fiber per acre per year than southern pine, and its pulp yield also compares favorably with that of southern pine and other softwoods. One emerging company, Kenaf Paper Manufacturing, Inc., is building a greenfield newsprint plant in south Texas. If this plant is a commercial success, more will follow.

What kenaf may be to the paper manufacturer, wheat straw could be to the board producer. North America produces huge quantities of residual wheat straw as a harvesting byproduct, with no identified use—until recently. North Dakota alone grows about 11 million acres of wheat each year, not to mention South Dakota, Montana, Kansas, and other states of the mid- and far west. The prairie regions of Canada grow even more. Until recently, the wheat straw was considered worthless debris. But wheat straw fiber has a composition similar to that of many commonly used softwood and hardwood species: 42% cellulose, 36% hemcellulose, and 22% lignin. At least one manufacturer has put it to use, making PrimeBoard, a straw-based particleboard made in Wahpeton, North Dakota, in response to the particleboard shortage during the early 1990s.

The plant was designed by a European firm experienced with agricultural fibers and board-making. A board line, which forms the wheat fiber into mats, feeds into a 7-opening, 5 × 18-foot Fjellman hot press. The resulting board complies with the ISO 9000 protocol and is well accepted in the trade for use in high-end laminated products such as furniture, doors, and millwork products.

Wood Fiber and the Future: A Glimpse Ahead

Wood products and paper businesses face an era of unprecedented change. As traditional raw wood sources dwindle and waste-wood quantities mount, we must redirect wood fiber away from the incinerator and the landfill—and into consumer products.

Searching for alternate raw materials, and products that use these raw materials, manufacturers will become product-driven rather than tree-specific. Alternatives such as kenaf, wheat straw, and urban waste wood generally cost less, appeal to the environmentally aware consumer, and offer competitive appearance and physical properties. The term "wood product" will come to describe a product classification rather than the raw material. New meanings, new materials, and new processes will reshape the forest industry in the new century.

Section Three:

The Mills: Adapting to Change

8 Engineered for Change: The Emerging Wood-Based Products

Engineered wood products (EWP) products are changing the way we think about wood and its uses in our daily lives. We demand style and beauty in our homes and commercial buildings, yet we also demand sensitivity to our environment. We demand more expansive structures, yet we demand provident use of our forest resources. Engineering and designing a structure that incorporates beauty, provident use of resources, environmental sensitivity, and affordable cost is a tall order.

There is a growing consensus that the world demand for wood can not be fully supported by solid sawn wood products. Fortunately, emerging technology offers solutions.

Engineered Wood: Improving Upon Nature

It is no secret that larger-diameter softwood logs are getting scarcer and more expensive. Old-growth timber, an industry mainstay in past years, is either nearly gone or locked up because of environmental concerns. What's left is the smaller-diameter second-growth softwood species, plus underutilized species, both softwood and hardwood. The producer now must create the products of the past, or their successors, from the trees of the present.

Engineered Wood Products (EWP) are manufactured by shaping and bonding together lumber, veneer, flakes, and other wood ingredients into functional, structural components. The resulting wood product is stronger and more versatile than the sum of its parts. Other beneficial features recommend the products—including light weight, ease of handling, and more economical installation. Some of these products, such as plywood, utilize centuries-old technology; others are just now emerging. Most have arrived on the scene within the last two decades.

Engineered wood products, rigorously engineered and tested to meet particular end-use specifications, are gaining wider acceptance as architects, specifiers, builders, and customers discover the benefits of cost, product performance, ready availability, and ease of handling. Consumers also like the fact that the raw material for EWP usually develops from abundant small-diameter timber, and that the engineering and product design process often requires up to a third less wood fiber to produce a comparably performing product.

There are two EWP industry segments: Structural Composite Lumber (SCL), which is a substitute for solid sawn lumber, timbers, and beams, and Engineered Panels.

Structural Composite Lumber

The SCL process provides a replacement for solid sawn wood by converting small-diameter logs—some as small as 4 to 5 inches (102–127 mm) in diameter—into billets and blanks that are far larger than anything available from other sources. SCL products are produced from these billets and blanks, or, in the case of glulam, from dimension lumber. Exhibit 8.1 lists the SCL product groupings.

Glulam

Glulam has been around for over a century; one of the earliest answers to the decline in availability of large trees, it was first used in 1893 to construct an auditorium in Basel, Switzerland. Currently, the product is gaining renewed popularity because of its versatility as a companion product with other engineered wood products. Glulam goes beyond the structural limitations of a huge solid timber; it has no internal weaknesses or tensions that could ultimately twist, warp, or split.

Glulam is to lumber what Laminated Veneer Lumber (LVL) is to veneer; the layers of lumber are laminated together in much the same way that veneer is laminated into LVL. The glulam beam is constructed with layers of wood, and adhesive-assembled in sequence.

The raw material used for glulam, dimension lumber, is sorted into lumber grades that were created specifically for glulam. These grade specifications are identified in published third-party grading rules, such as paragraphs 153 and 154 of *Standard No. 17, Grading Rules for West Coast Lumber.* The West Coast Grade Rule is specifically

Exhibit 8.1: Structural Composite Lumber

- Glulam (GL)
- Laminated Veneer Lumber (LVL)
- I-Joists
- Parallel Strand Lumber (PSL)
- Laminated Strand Lumber (LSL)
- Other lumber-type products
    - Engineered Strand Lumber (ESL)
    - Scrimber
    - Star Cutting
    - Wood/Plastic Fiber Composite (Trex)

engineered for the species in the Pacific Northwest region of the United States; other regions and countries have their own specific grades.

The grades are more specific and demanding as to wood species, wood characteristics, moisture content, and visual specifications than the usual dimension lumber grades. Moisture content, usually limited to 19% on kiln-dried (KD) lumber, trends down to 13% to 15% in lamstock. The lower moisture content (MC) is ideal for gluing, and approximates the equilibrium MC in service. The lower MC also prevents surface checking when the product is in service.

Glulam is a versatile structural product. It can be manufactured into stock sizes for headers and joists, or in long-span structural members such as parabolic arches or bowstring trusses. Manufactured into an almost endless variety of designs, glulam serves the architect both in hidden applications and where appearance is important. (It is said that glulam gives the architect "artistic freedom without sacrificing structural requirements.")

Although glulam production fell about 2% in 1997 in North America—to 315 million board feet— the decline was caused by the economic problems in Japan, a leading export market for this product. Production is expected to rebound to about 337 million board feet by the year 2002, a 7% gain.

Laminated Veneer Lumber (LVL)

Laminated Veneer Lumber (LVL) is a layered composite of wood veneers and adhesive similar to glulam. First used during World War II to make airplane propellers, it is one of the latest in a growing number of wood composites designed for structural applications. LVL manufacturing is an established industry in both the United States and Europe, and it's making significant headway in Asia for both structural and nonstructural applications. For example, the Japanese Ministry of Agriculture, Forestry and Fisheries recently opened the door to the manufacturing and consumption of structural LVL in Japan through the promulgation of Notification No. 1443, the Japanese Agricultural Standard for Structural Laminated Veneer Lumber.

Trus Joist International's founders, Harold Thomas and Art Troutner, are credited with inventing the modern LVL industry. Their 1960s quest for lumber with suitable characteristics for prefabricated I-joist flange material prompted the invention of structural LVL. Troutner and Thomas created a structural LVL light framing replacement that is strong, lightweight, and uniform in structural properties.

The major differences between LVL and the older product, plywood, are (1) LVL's need for higher-strength veneers, (2) the parallel rather than perpendicular placement of the veneer, (3) the thicker and larger size of the LVL assembly, and (4) the precise quality control testing and auditing procedures used in LVL manufacture.

The species and log types used in LVL depend upon the visual and structural properties required in the finished product. Sought-after species include Douglas fir (*Pseudotsuga menziesii*), hemlock (*Tsuga heterophylla*), larch (*Larix occidentalis*), and the southern pines, such as loblolly (*Pinus taeda*), longleaf (*Pinus palustris*), and slash pine (*Pinus elliottii*). However, manufacturers use other species in certain areas where they are cheaper and more available, such as yellow poplar (*Liriodendron tulipifera*), aspen (*Populus tremuloides*), eucalyptus (various species), and radiata pine (*Pinus radiata*).

Ring count and growth site largely determine structural strength characteristics within a species. Generally, a higher specific density and ring count, a greater proportion of summerwood, and a more fertile—yet dry—growing site will translate to higher structural properties in the resulting veneer.

For example, the suppressed understory of a Douglas fir forest has trees with a high ring count. These high-ring-count trees will yield the highest proportion of veneer suitable for LVL from a second-growth stand. The procurement forester uses an increment borer to determine the ring count of a standing tree—a good indicator of the relative structural characteristics of a tree and the stand.

The manufacturing process begins with the rotary peeling of logs, some as small as 6 inches in diameter. The veneer is clipped to the widest useable width at the green clipper, then stacked for transportation to the downstream veneer dryer. The veneer passes through a high-temperature dryer that reduces the moisture content to 7% or less. The dry veneer is then both visually and mechanically graded at the dryer outfeed.

The veneer passes through an ultrasonic grading device, which makes a nondestructive evaluation of strength characteristics. The device measures the ultrasonic response through the length of the sheet as the two veneer ends pass through a pressure sensor simultaneously. Inspectors check the marked veneer for knots and knothole sizes, splits, voids, and other physical defects; these must not exceed a specific quality standard.

All veneer is placed with the grain parallel in the rough assembly; this enhances the natural strength properties along the length of the plies. The LVL assembly, now called a billet, is curtain-coated with an adhesive, usually a PF resin mix. The billet feeds into a hot press, where heat and pressure consolidate and bond the plies. The wood species, the number and thickness of the individual graded veneer sheets, their placement within the billet, the relative compression at the hot press—all these factors determine the finished LVL's relative structural strength.

In marked contrast to the relatively small size of plywood (usually 4 × 8 feet or its metric equivalent), the resulting LVL billet is 2 to 4 feet in width and 60 feet or more in length. The longest length manufactured is governed by the length of the hot press or, in a continuous press, the ability to process or transport the long lengths. LVL is manufactured in thicknesses from 3/4 inch (19 mm) to 2 1/2 inches (64 mm) and more; at least one manufacturer offers a 3 1/2 inch (89mm) thickness

Laminated veneer lumber shares common attributes with plywood and glulam. Each incorporates layers of wood and veneer, with plies placed according to a "recipe," with an adhesive bonding the plies or layers into a billet or assembly. LVL has a variety of end uses; some overlap the markets for other structural composite lumber (SCL) products, including glulam, parallel strand lumber (PSL), and oriented strand lumber (OSL).

Engineered wood products are increasingly manufactured from small logs.

The LVL products usually cost more than solid sawn lumber; but offer attributes generally lacking or increasingly unavailable in solid wood. Modern builders increasingly seek out LVL for its high reliability and lower variability. LVL works well for scaffold plank, concrete forming material, transmission cross arms, and truck bed decking with hardwood face veneers.

Prefabricated Wood I-Joists

Prefabricated wood I-joists and I beams form a fast-growing segment of the engineered wood product industry (see Table 8.1). Recent years have seen annual production increases of 25% and more. In 1997 North American wood I-joist production rose to a record 627 million linear feet, nearly 26% more than the previous year's 498 million. Nearly one-third of America's residential wood flooring systems used I-joists during the 1997 construction season. The industry expects jumps of 50% to 60% within 5 to 7 years. APA–The Engineered Wood Association expects this trend to continue well into the 21st century and spread to other countries.

Major U.S. manufacturers of I-joists include Trus Joist MacMillan (TJM), Willamette Industries, Louisiana Pacific, and Boise Cascade. In Europe, Vanerply AB is setting the pace. This Swedish company already has code approval in four northern European countries—Denmark, Sweden, Norway, and Finland—and is seeking approval for the U.S. and the U.K. Other companies in other countries are expected to follow suit.

Table 8.1: U.S. and Canadian Wood I-Joist Production

1995–1997 Actual, 1998–2002 Estimated (in million linear feet)

	1995	1996	1997	1998	1999	2000	2001	2002
U.S. Production	358	444	547	680	775	855	950	1,000
Canadian Production	39	54	80	105	120	140	145	150
North American Total	397	498	627	785	895	995	1,095	1,150

Source: APA–The Engineered Wood Association, as reported in Crows, October 1998.

Wood I-joists are manufactured by gluing a top and bottom solid-sawn lumber or LVL flange to a web of either plywood or oriented strand board (OSB). The flange material, sawn to the desired width, is joined into long lengths as the continuous I-joist is assembled. A groove machined on the inner side of each flange accommodates the placement of the web material.

The web component, cut to the precise width to match the required distances between flanges, is seated and glued as the top and bottom flanges are pressed onto the web material. The flanges and the web, matched together, form a completed I-member when cut to a specified length and post-conditioned. The "I" configuration provides a high strength-to-weight ratio.

Wood I-joists are a popular substitute for solid lumber framing in floor systems because they can span longer distances and provide greater design flexibility as well as high bending strength and stiffness characteristics. The uniform stiffness, strength, and light weight of this prefabricated structural product make it well suited for floor joists and rafters for both residential and commercial construction.

Production of I-joists, glulam, and LVL trends higher each year (see Table 8.2). The production of I-joists and LVL was expected to triple between 1993 and 2003, with steady growth predicted for glulam.

Parallel Strand Lumber (PSL)

Parallel Strand Lumber (PSL) is another veneer-based product. While billets of LVL or faces and backs of plywood call for full veneer sheets—or "54's" as they are often called—the damaged or high-defect full sheets make prime raw material for PSL. Producers first dry this veneer and chop it into long strands to remove the defects. Next, they apply an exterior glue such as a PF adhesive mix to the dry strands. With the strands oriented to the length of the member, the material is formed into a continuous assembly, then compressed under heat and pressure into a rough billet.

The rough billet dimensions are typically 11 × 16 inches (280 × 406 mm) in cross section, with the length designated by customer order. The billets are ripped into timbers, beams, and columns for post and beam construction, or ripped into beams, headers, and lintels for residential light-frame construction. PSL is used in both exposed and

Prefabricated wood I-joists are replacing wide-width lumber.

Table 8.2: North American Engineered Wood Production

1980–1997 Actual, 1998–2002 Estimated

	Glulam (MMBF)	I-Joist (MM linear ft.)	LVL (MM cubic ft.)
1980	214	50	3
1981	199	50	4
1982	172	60	4
1983	201	70	5
1984	240	80	5
1985	258	100	7
1986	346	110	8
1987	292	120	9
1988	312	120	11
1989	337	130	12
1990	339	135	16
1991	277	175	18
1992	270	280	20
1993	250	398	25
1994	276	422	28
1995	295	397	28
1996	322	498	32
1997	315	627	38
1998	285	785	46
1999	286	895	54
2000	297	995	62
2001	318	1095	70
2002	337	1150	77

Source: APA–The Engineered Wood Association, as reported in Crows, October 1998.

Note: Data for years prior to 1994 are estimates from various articles and reports. The APA began mill surveys and estimates in 1994.

Prefabricated wood I-joists, wrapped and ready to ship.

hidden applications, such as intermediate and large members in commercial building construction.

Originally a product of MacMillan Bloedel's R&D effort, PSL was further refined by a later Trus Joist MacMillan (TJM) partnership. PSL, a proprietary product, is currently manufactured under the trade name Parallam at two locations in the southeastern U.S. and at the original factory in the lower British Columbia mainland.

Oriented Strand Lumber (OSL)

Oriented Strand Lumber (OSL) (also known as Laminated Strand Lumber or LSL), another TJM proprietary product, is more akin to Oriented Strand Board (OSB) technology than to veneer. For OSL, a whole log is flaked into strands rather than peeled.

Manufacturers start by debarking and processing the log through the rotating knives of a stranding machine, producing 12-inch (305-mm) lengths. The resulting strands are much longer than the 3- to 6-inch (75- to 153-mm) strands of OSB, but the thickness is about the same. Off-tolerance strands are screened out before billet assembly and used as boiler fuel or for other uses.

The stands are dried in a rotating drum and adhesive is sprayed onto the fiber. Parallel-orienting the strands during billet formation provides greater strength parallel to the length of the finished product. The billets are adhesive-cured by a stationary steam injection press in the assembly. Billets are manufactured with dimensions up to 5.5 inches (140 mm) thick and 8 feet (2440 mm) wide, up to the length of the press platen, or in longer lengths with a continuous press.

The longer strand length, the screening out of the undesirable strands, the parallel orientation of the strands, and the densification of the fiber give OSL its superior strength. OSL is also successful because it can incorporate many timber species and types. Aspen, poplar, and other relatively low-density species can be used, along with small or defective logs unsuitable for peeling or lumber manufacturing. Currently two TJM plants manufacture OSL, one in Deerwood, Minnesota, the other in Hazard, Kentucky. The product is marketed as Timberstrand LSL in North America and Intrallam in Europe. It is often used in residential construction as rimboards, and a host of other structural and nonstructural applications are being developed.

Other Lumber-Type Products

Weyerhaeuser is producing Engineered Strand Lumber (ESL), a lumber product created with strands of fiber layered and bonded together. Specifically designed as a replacement for both lumber and plywood in upholstered furniture frames, it is a versatile product used for arms, backs, and other parts. Strandwood Molding manufactures a similar product, in which flaked aspen and polyurethane adhesives create a lumber/panel material used as a wood-based seating component. Scrimber Lumber is another innovative process.

Scrimber is a whole-log technology formerly used by Scrimber International of Mount Gambier, South Australia. Once the log is debarked, the fiber is crushed lengthwise into bindles of interconnected and aligned strands that maintain the original orientation of the grain. The scrimmed mat is dried, adhesive is added, and the assembly is formed into the desired shape and pressed while the adhesive is cured by radio-frequency energy. Georgia-Pacific is currently evaluating this technology for turning southern pine thinnings into a value-added product.

Another SCL innovation is star cutting. This new way to break down a log into lumber was developed at the Royal College of Technology (KTH) in Stockholm. The log is sawn at a 90-degree angle from the surface inwards toward the pith so that the growth rings are vertical to the surface of the developing lumber.

The technology is being commercialized by Nova Wood, a subsidiary of the Swedish firm SCA. Nova has built a plant in Tagsjjoberg, Sweden. The triangular lengths will be made into a knot-free laminate called Primfog at another plant situated near the Nova mill. The product is said to be an improvement on conventional lumber, with a higher log yield, a harder finished lumber surface, more uniform seasoning, and fewer splits in service.

Trex lumber uses wood fiber and consumer plastic waste as its primary raw materials. Trex's appeal lies in its raw material sources, its finished appearance, and its low maintenance requirements in exposed applications. It puts to use hardwood waste such as sawdust and other wood residuals, plus reclaimed plastic such as shopping bags and stretch film—materials that would otherwise end up in the incinerator or landfill. Trex is exceptionally resistant to moisture, rot, and insects—it resists termites and other

I-joists, glulam, and OSB—an EWP Flooring System.

wood-eating pests without the use of toxic chemicals. It can be sawn, fastened, drilled, routed, and sanded like wood.

Some consider its wood-polymer composition ideal for decking or other exposed applications. However, Trex is not to be used as a structural member such as a post, beam, or joist because it is both heavier and more flexible than wood. Its dimensional stability in service through weather cycles is still being evaluated.

Trex lumber is a start of a trend: lumber products that are something more than solid wood and wood fiber. The newer products will marry wood with other materials to gain the advantages of both. Builder and consumer will need to be more knowledgeable and look further when selecting products for a project. With more and more innovative products coming to market, consumers will no longer have to adapt a product that's almost right for their needs.

Engineered Structural Panels (ESP)

Worldwide consumption of structural and other panels is huge—and growing. Panels, designed to meet structural or nonstructural-based specifications, satisfy a variety of applications, from display tables in an Istanbul street market, to a water-bed platform in California, to a treated all-weather wood foundation in Texas. ESP panels include three major categories (see Exhibit 8.2), and other panel types are being developed. The uses for panels are almost endless; cost, utility, and availability drive the demand.

Exhibit 8.2: Engineered Structural Panels

- Plywood
- Comply
- Oriented Strand Board (OSB)
- Other boards

Plywood

Plywood has been used in decorative objects and furniture since the pharaohs ruled Egypt 3,000 years ago. But plywood did not debut as an engineered wood product until Portland Manufacturing Company displayed its panels at the 1905 Lewis and Clark Centennial Exposition.

Hundreds of uses soon followed. Doors, furniture, trunks, and crates were only a few of the early applications for which plywood's size and dimensional stability offered a clear advantage. By mid-century, softwood plywood had become the sheathing of choice for residential and commercial construction. When its original raw material, old-growth Douglas fir, became a casualty of overcutting and set-asides, producers turned to other softwoods with diameters as small as 6 inches.

Some think that consumers will soon abandon this once-pioneering product in favor of the newer engineered wood products. However, the versatility of plywood's raw material, the flexibility of the manufacturing process, and its appearance and physical properties assure its future in EWP applications. Many will still rely on plywood's weight/strength ratio, dimensional stability, and tested performance in major weather and seismic events. Pound for pound, its stiffness and racking mechanical properties outclass the competition, both wood and nonwood.

Comply

Comply, a mid-20th-century variant of softwood plywood, combined particleboard and plywood technology to create a new panel type, with a face and back of veneer and a particleboard-type core. An extra layer of veneer in the center of the thicker comply panels gives added racking strength and dimensional stability.

Only a few manufacturers currently manufacture the comply panel. Comply's relatively high manufacturing cost, its heavier weight in comparison to plywood (which means higher shipping costs), and its continuing price competition with plywood and the newer composite boards, such as OSB, have relegated the product to a minor role in the marketplace. However, many believe it's a superior underlayment panel for residential floor systems. Nailed and glued to wood I-Joists, comply forms a system reputed to be the premier silent floor. Comply technology is also used for roof and wall sheathing, and occasionally for floor joists and studs.

Plywood is the oldest EWP flooring system.

Oriented Strand Board (OSB)

Oriented Strand Board (OSB) is a second-generation panel originating from the waferboard technology developed in 1954 by Dr. James d'Arcy Clarke, an internationally renowned wood scientist. Dr. Clarke was an environmentalist with practical vision. He knew that big peelers were scarce and expensive, and that underutilized weed trees were already naturally regenerating in many regions of North America. Wafers from a variety of these aspen and poplar species—when properly dried, bonded together with a PF resin, and hot pressed into a flat panel—had many of the properties of structural plywood.

Others took the concept further to produce a board with the core wafers perpendicular to the face layers, designated Oriented Stand Board (OSB). The resulting improvement in strength and performance characteristics made it a better plywood substitute.

The new OSB structural properties, and its recognition by American manufacturers such as Potlatch, Louisana Pacific, and others in the late 1970s, soon led to additional testing and development in fiber alignment, adhesive development, and processing innovations. The Engineered Wood Association (APA) began working on code approvals, design information, and other promotion activities that led to remarkable industry growth. Starting with little or no market share in the early eighties, OSB will have captured half or more of the North American structural panel market by the turn of the century (see Exhibit 8.3).

Exhibit 8.3: North American Panel Production

1990–1997 Actual, 1998–2002 Estimated (in million square feet, 3/8 basis)

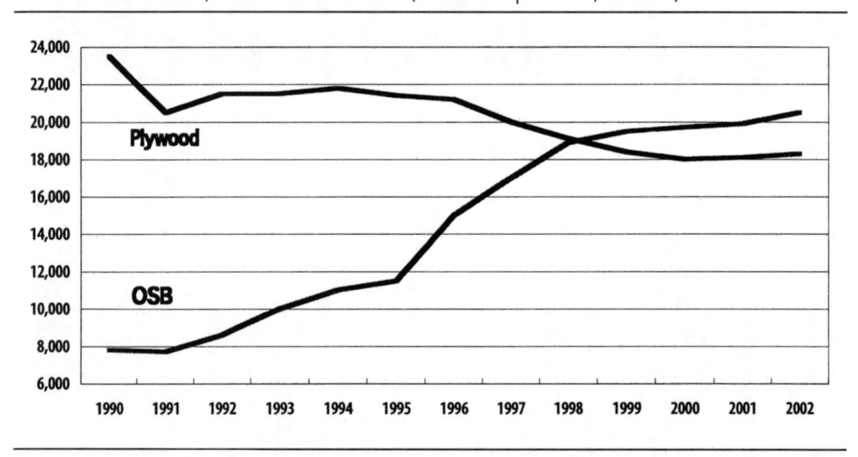

Source: APA–The Engineered Wood Association, as reported in Crows, August 1998.

While the acceleration in market growth can be expected to level out over time, it's clear that the North American panel innovation has gone international, with a Republic of Ireland plant the first of many to come. OSB will continue to be a growth industry, with burgeoning international manufacturing capacity and consumption.

The Other Boards

Triboard, manufactured by Juken Nissho Ltd., a Japanese-owned, New Zealand–based forestry company, is a wood composite panel that integrates Medium Density Fiberboard (MDF) and OSB technology. The panel is constructed from a core of strands sandwiched between surface layers of fiber. Steam injection pressing allows the construction of a thickness in excess of 4 inches (100 mm).

Designed for walls in residential housing and apartment buildings, Triboard is reportedly rated for 90-minute firewall protection. Triboard serves well for both interior and exterior walls. The manufacturer claims that this panel can replace traditional 2 × 4 framing. The Japanese are using it to supercede post and beam construction.

Juken Nissho is exporting the panel to Japan for other uses: pallet bases, basketball court floors, squash court walls and floors, and professional table-tennis tables, as well as worktops, solid doors, stairs, shelves, and flooring.

The Triboard panel is one example of the engineered wood-based boards that will be developed for structural applications in the years ahead.

Engineered Wood Products and Company Performance

To some manufacturers of plywood and solid sawn lumber, engineered wood products are still the "new kid on the block," but they like what EWP has to offer. Compared to solid sawn lumber or plywood, the newer EWP technologies use more of the log. And the log is usually a lower quality and smaller diameter than the raw material it replaces.

In addition, the raw material price often compares favorably with lower-cost paper mill furnish. The raw material can also be manufactured into a wide variety of products. Finally, EWP prices are more stable and often higher than the prices of their solid wood or paper counterparts.

The fact is, companies adopting the newer EWP technologies are improving their financial performance—and those that aren't are not.

EWP-focused companies reap improved earnings as these products gain a higher percentage of their sales. However, just manufacturing EWP is no guarantee of sustained superior results. Engineered wood products demand a higher level of technical and management skills than the more traditional wood products. These newer products have superior margins, but they require intensive research, development, field testing, and manufacturing control. A market entrant is well advised to establish a product development and EWP manufacturing culture well in advance of making and delivering product to the field.

Overall, customers are becoming increasingly receptive to the newer engineered wood products, both because of the cost and as solid sawn counterparts become more scarce. Builders are not happy with the lower-quality second-growth lumber and plywood products they see coming into the market. In the face of these concerns, coupled with the forestland lockups and the growing environmental sensitivity of the consumer, both the builder and the consumer are drawn to the newer products. Engineered wood products will find an even warmer welcome as the forest industry moves forward into the 21st century.

Adhesive Systems and Wood Products

9

Wood-bonding technology uses organic and inorganic adhesive systems to provide the linkage between the forest products of the past and those of the present and future. Adhesive systems enable the shaping and gluing of wafers, flakes, veneer segments, lumber components, and other wood ingredients into attractive and functional products. The forest-based industry will find it essential to understand adhesives, their use, and the emerging gluing innovations as the industry progresses into the new millennium.

Improving the Use of the Wood Resource

Wood and glue have been woodworking companions for centuries. But in the last decades of the 20th century, innovators are developing new—and even exciting—applications for adhesives at an unprecedented rate. Small-diameter trees, with short-rotation harvest cycles of one, two, three, or four decades, are the wood furnish of the present and the future.

Just as wheat is the basic ingredient of many foods, these short-rotation trees become the raw material for a growing variety of products with specific attributes. Wood-bonding technology is the key to making use of these trees.

In the mill, the first-stage products of a tree are now something more than lumber, veneer, and chips; products now include flakes, wafers, strands, and fiber bundles, which become panels, lumber, and massive wood components. Adhesive-dependent secondary manufacturing then completes the transformation into finished goods.

Even lumber and veneer have become something more than the traditional solid wood products. Narrow-dimension lumber becomes headers, beams, and massive trusses. Veneer becomes huge sheets of laminated veneer lumber (LVL) or cants (Parallam). The latter depend on veneer as a primary ingredient; no less important is the resin system, making big components out of small ones.

This fundamental change is shifting the log breakdown process from big pieces to little segments, which are then reconstituted into the products we want. The desired attributes are now engineered into the manufacturing process.

Adhesive technology can also enhance the beauty and utility of wood, even when compared to its big-tree ancestors. The unique structure of an I-joist composed of small-log–sourced ingredients of veneer and wafers needs one-third or less fiber for a similar structural application when compared to its solid wood competitor. Glulam assemblies of narrow-dimension lumber benefit from layers of fiber-reinforced plastic that take the place of more wood or wood with lower structural attributes.

Then there are the organic and inorganic overlays, which add beauty and value to a basic surface. Dimension lumber becomes a door stile with vertical grain (VG) characteristics. Oriented Strand Board (OSB) becomes a soffit or exterior lapsiding. A flat sheet of Medium Density Fiberboard (MDF) becomes a lookalike for expensive rainforest hardwood lumber. Designers and builders now choose from a varied palette of substitutes for the look of solid woods of the past, using overlay materials such as the melamines, phenolic films, thermoplastics, high-pressure laminates, metal foils, fiberglass, textiles, and wood veneers as thin as 1/42 inch (0.61 mm). The resulting products often offer more durability, dimensional stability, and beauty—at a lower cost—than the products of the past. Again, adhesives are the key; they allow for freedom of design and economy of choice.

Wood Bonding: The Theory

Wood bonding and adhesives make up one of the most widely used—but least understood—technologies in the forest products sector. Many who struggle to master the science of wood bonding find it too highly technical. The reader who feels compelled to understand the cause-and-effect relationship between bonding and the substrate being bonded may look to the work of A. Pizzi in *Advanced Wood Adhesives Technology.* This author outlines the phenomenon of adhesion by describing five theories: the mechanical entanglement/interlocking theory, the diffusion theory, the electronic theory, the adsorption/specific adhesion theory, and the covalent chemical bonding theory.

In the face of such complexity, most wood-products practitioners may want to leave the theory to the lab and the practical application to the mill. But the growing choice of substrates, adhesive systems, and products demands that industry participants have at least a basic understanding of the theory of bonding.

In simplest terms, adhesives are liquids that convert to solids, or solids that convert to liquids and then reconvert to solids, during the process of bonding adjacent surfaces (called substrates). Bond formation requires the adhesive to attain a close conformation between surfaces. Some have demonstrated that if adjacent surfaces are smooth enough when compressed together, the inherent molecular attractive forces of the substrates will come into operation with no adhesive necessary.

In the case of wood, the adhesive compensates for both the inherent porosity of the material and the added roughness created by machining and preparation.

Adhesives undergo at least four distinct actions in establishing a bond between the adjacent surfaces:

- A glueline is formed as the adhesive flows or is applied across the surface and into the high and low areas prior to pressure application.
- The adhesive is often transferred to the opposite surface, as in a printing operation, during gluing, such as in the plywood layup process.
- The adhesive penetrates the wood pores and interstitial structure of the substrate.
- Solidification takes place, producing strength and bond durability as the substrates remain in intimate contact during curing of the glueline.

One author succinctly describes the process of adhesion as follows:

> Physical adhesion is generally considered to be the main mechanism by which adhesives attach to wood. However, mechanical or interlocking actions always occur and contribute their measure of strength. In fact, it is not unreasonable to state that the attraction between molecules can also result in a maximum of interlocking and coincidentally a maximum of strength. (Marra 1992)

Three primary adhesive mechanisms take place during the bonding process. First, mechanical bonding occurs as the tendrils of glue penetrate the pores and interstitial structure of the adjacent substrates (similar to applying plaster on a wall). Second, a chemical reaction occurs between the adhesive molecules and wood molecules to form a new compound. Third, a physical bond is created by molecular attraction, such as occurs in magnets.

Whatever the underlying theory, a growing body of knowledge, experience, and additional research and development efforts are producing more durable and product-specific bonding materials.

Modern Adhesive Types

The steady growth of adhesive consumption in North America keeps pace with the growth of wood-adhesive composites. For example, North American (excluding Central America) consumption of adhesive resin solids, including an array of primary glued products, was 1.50 million metric tons in 1994; in 1997, 1.65 million metric tons were consumed, a net annual increase of 3.7%.

However, according to figures published by the United Nation's FAO 1994 report, North American consumption accounted for only 28% of world production of plywood and composites. The adhesive business is a large and important industrial sector; its importance can be further tested by measuring its cost as a proportion of a product's total manufactured cost. In some products the cost of resin is a relatively small portion; in others, resin accounts for 32% or more of the cost.

The North American 1997 adhesive volume was consumed into 56.9 million cubic meters of wood composites such as plywood and LVL, OSB, particleboard, MDF, and other products (see Table 9.1). Consumption of three of these products (LVL, OSB, and MDF) is experiencing exponential growth around the world.

Table 9.1: 1997 North American Wood Composites

Industry category	Percent	Quantity (million m^3)
Plywood, LVL	35	20.107
Oriented Strand Products	26	15.017
Particleboard	18	10.180
Medium-Density Fiberboard	6	3.163
Other	15	8.422
TOTAL	100	56.889

Source: Terry Sellers Jr., "Adhesive Industry Matching Wood Composite Needs." *Panel World*, September 1998, p. 12.

Table 9.2: Estimated 1997 Consumption of Wood-Based Adhesives in North America

Polymer type	Metric tons	Percent
Phenolics	567	34.3
Amino	980	59.3
Isocyanate	50	3.0
Polyvinyl	47	2.8
Resorcinol types	<5	0.3
Soy-modified casein	5	0.3
TOTAL	1654	100

Source: Terry Sellers Jr., "Adhesive Industry Matching Wood Composite Needs." *Panel World*, September 1998, p. 12.

The 1997 estimated consumption of major industrial wood-based adhesives in North America is illustrated in Table 9.2. Amino and phenolic polymer-type adhesives account for over 90% of total consumption.

> More than 59% was amino resins (urea-formaldehyde [UF], melamine-formaldehyde [MF], etc.). Approximately 34% of the volume was phenol-formaldehyde (PF)-type resins (includes resorcinol-formaldehyde [RF] types), and the balance 6% was divided among a variety of other synthetic binders such as polyvinyl acetate (PVAc), cross-linked vinyl (X-PVAc), soy-modified casein, and polymeric diphenyl methylene diisocyanate (PMDI). In addition, a vast and increasing volume of natural-based and synthetic binders are in paper products,

overlays and surface coating industries, and cement/gypsum-type binders incorporated with wood fiber exist. (Sellers 1998)

A comprehensive description of the most widely used adhesives (Table 9.3) follows.

- **Urea-formaldehyde resin (UF).** This adhesive system is versatile and low-cost because its principal raw materials, urea and formaldehyde, do not contain benzene or aromatic compounds. The urea-formaldehyde resins (acid-curing and heat-accelerated) are widely used in the manufacture of hardwood and softwood panels and other veneer-based products. Manufacturers rely on these adhesives' characteristic attributes: fast cure and high strength. But they have a weakness: low resistance to high temperatures and moisture over the long term. UF resin-based adhesives will not withstand continuous cycles of wetting and drying and will begin to delaminate at about 140°F (60°C) and 60% relative humidity. Panels glued with these adhesive types should not be used where they would be exposed to weather or elevated temperatures.
- **Melamine-formaldehyde resins (MF).** Melamine-formaldehyde (MF) resins are amino-aldehyde products. They are more durable than the UF adhesives, but more costly to manufacture. UF/MF adhesive blends present a cost and performance compromise. These blends are used in hardwood plywood and laminating industries for applications that require improved durability while maintaining light-colored gluelines.

  In their spray-dried form, MF resins are more costly to manufacture than UF resins. They are more expensive on a solids basis than either UF or most liquefied phenol-formaldehyde (PF) resins. However, it's possible to make large quantities of MF resins as cheaply as PF resins. While an MF resin is an improvement over a UF resin, it is even less durable under wet, humid, or high-temperature conditions.
- **Phenol-formaldehyde resin (PF).** PF resin adhesives are the principle resins used in North America for bonding structural plywood and OSB. This resin type is also used to bond dense hardwood plywood. When properly cured, PF adhesives are waterproof and often more durable than the wood itself. Less desirable features are higher cost, a dark-colored glueline, and its requirement for a lower veneer moisture content during the gluing process.

  However, the superior durability of the PF adhesive far outweighs its disadvantages in a wide array of applications. The phenolic resins are produced under alkaline conditions (pH in excess of 7) using molar ratios of 1.6–1 to 2.5–l, formaldehyde to phenol. The excess formaldehyde facilitates a cross-linking density that results in an excellent moisture resistance in the cured state, low flammability, high tensile strengths, and good dimensional stability.

  The raw materials for PF resins are petroleum-based. Natural gas is the feedstock for the formaldehyde ingredient, while crude oil supplies the phenol portion. These raw materials and their alternate uses expose PF resin formulations to price

Table 9.3: Commonly Used Adhesives

Adhesive resin	Characteristics	Typical application
Urea-formaldehyde (UF)	Hot- and cold-setting; acid-curing with heat and/or catalyst-accelerated fast cure; cold-water resistant; colorless; may emit formaldehyde in use.	HW flooring, type II plywood (decorative), particleboard, fiberboard; interior exposure.
Phenol-formaldehyde (PF)	Hot-setting; normally cured above 220°F; usually highly alkaline for rapid cure; waterproof; dark in color.	Structural plywood; truss components, OSB, and waferboard; exterior exposure.
Melamine/urea (MF/UF)	Hot-setting; heat and catalyst accelerate cure; warm-water resistant; colorless.	Plywood (decorative), flatbed stock, end joints in laminating; interior and limited exterior exposure.
Emulsion polymer/isocyanates (EPI)	Cold- or hot-setting; two-component system; room-temperature curable; water- and temperature-resistant; neutral in color; non-formaldehyde-emitting.	Laminating wood to wood and wood to nonwood substances, millwork contact, and pressure-sensitive; interior and exterior exposure.
Isocyanates (MDI)	Hot-setting; water and heat accelerate cure, waterproof under severe conditions; neutral in color (press release agent tans wood surfaces).	Waferboard, OSB, and particleboard; interior and exterior exposure.
Melamine-formaldehyde (MF)	Hot-setting; heat and catalyst accelerate cure; water resistant; colorless; shipped in spray-dried form.	Laminated beams (moderate use in U.S.), end joints in laminated truck decking, and Type I HW plywood (decorative); limited exterior exposure.
Polyvinyl acetate and derivatives (PVAc)	*Homopolymers* (compounded or uncompounded): cold-setting; somewhat flexible when cured; poor water resistance; light in color; tendency to creep under load; gap filling; rapid tack.	Assembly gluing, gluing face veneers to core stock, and edge banding; interior exposure.
	Coplolymers: cold- or hot- (and Rf) setting; heat and catalyst accelerates cure; moderately water resistant (cross-linked); more rigid and gap filling.	Edge and face stock, HW plywood, and finger joints; interior and limited exterior exposure.
Phenol-resorcinol formaldehyde (PRF)	Room-temperature- and warm-setting; heat and catalyst accelerate cure; waterproof under severe conditions; dark in color; particularly suited for difficult bonding conditions.	Bridge and pier components, laminated beams, and truck decking; interior and exterior exposure.
Resorcinol-formaldehyde (RF)	Cold- or hot-setting; cure accelerated by catalyst and heat; waterproof under severe conditions; dark in color; particularly suited for difficult bonding conditions.	Laminates, ship components, outdoor furniture, and fire-rated panels; extreme exterior service.

Source: *USFS General Technical Report 50-71* (1988).

pressures as demand shifts to products such as polyethylene or to more common products such as gasoline diesel, lubricants, and asphalt.

- **Isocyanates.** Adhesives based on polymethylene polyphenyl isocyanate and methylene bisphenyl diisocyanate (MDI) have high bonding strength, are formaldehyde-free, and can be cured at either room or elevated temperatures. This adhesive type was introduced into the softwood plywood industry as the second part of a two-part isocyanate/PF combination that is used to successfully bond high-moisture-content southern pine veneer.

  The use of isocyanate-based adhesives has been restricted by more than the high cost of the ingredients. Health hazard concerns and sticking problems—such as panels glued to the hot press platens—have discouraged its wide usage.
- **Polyvinyl acetate (PVA).** Polyvinyl acetate (PVA), a versatile thermoplastic polymer, comes in many forms and is suitable for bonding a variety of organic and inorganic substrates. Most often, the PVA-type adhesives are compounded formulations in emulsion form. Ingredients used in the formulation include polyvinyl alcohol, cross-linked phenolic or isocyanate—the latter as thermosets—plus wood flour, kaolin, and other plastic additives, and such as butyl phthalate or diisobutyl phthalate. Manufacturers use the resulting adhesive in a wide variety of wood products, including doors, furniture, millwork, and windows.

These adhesives, plus a growing variety of others, are the industry workhorses. And researchers in the field and the laboratory constantly turn out new combinations with the addition of other polymers and ingredients, striving to meet the demand for improved and more versatile adhesive systems.

What's Ahead

Researchers now combine high-tech computer applications—rate-dependent durability modeling, molecular modeling, finite element analysis—with more traditional empirical evaluation methods to unlock the secrets of adhesive technology. Computer capabilities benefit both the older and newer adhesive formulations. No less importantly, improved analytical equipment and measurement procedures allow developers to fine-tune adhesive properties.

The search for improved job-specific adhesive properties is being matched by a search for improved processing equipment, such as continuous production lines, innovative adhesive applicators, new methods for curing the glueline, and ways to comply with environmental constraints, just to name a few. Adhesive suppliers must develop ever more high-performance, highly targeted adhesive formulas to meet the demands of a wood products industry with faster and more efficient equipment and highly specific applications. They are responding with these recent developments:

- **Faster cycles and higher moisture content (MC) substrate bonding.** The two-part adhesive (PF/isocyanate) used in high-MC plywood layup is old news. A newer

Modern adhesives are essential in the panel-making process.

development is the two-part gluing of OSB using an isocyanate-type adhesive as core material. Researchers are working to develop an adhesive system without the health risks, short pot life, and sticking problems of the older isocyanate-based formulations. Both manufacturer and supplier see the need for adhesives with ever-shorter curing times and tolerance for higher substrate MC. Running a composite board plant with an initial $80 to $100 million capital price tag tends to create an incentive to run faster with lower processing costs. Faster-reacting adhesives now under development may include either a modified isocyanate polymer or a catalyst addition.

- **Green fingerjointed lumber adhesive.** The Greenweld process, a New Zealand innovation, uses a melamine-urea resin and a hardener containing resorcinol in combination with a curing agent. This allows the fingerjointing of green lumber, usually short lengths, into specified lengths. This process has the potential to materially increase the yield of useable lumber from the log.
- **Phenol formaldehyde (PF) one-component adhesive.** Makers of veneer-based parallel strand lumber (PSL) and oriented strand lumber (OSL) now utilize a PF/wax combination as one component of an adhesive system technology to provide more uniform and efficient application of the adhesive during the process.
- **Reducing thickness swell and increasing strength adhesive.** Researchers in the U.S. and other countries are tackling the problem of thickness swelling in OSB. Their goal: an adhesive formulation to both further reduce swelling and increase the composite board's strength.

These developments are just a few examples of how public and private development work all over the globe is enhancing the performance of traditional and newer adhesive systems. Manufacturing properties and long-term performance aren't the only issues; environmental concerns also drive the search.

Chief among these, particularly in the United States, are the challenges of life-cycle durability and emissions reduction. Simply stated, wood is a carbon sink; the longer a wood product lasts, the longer the carbon sink remains undisturbed and the fewer the trees that need to be harvested—or so the thinking goes. The consumer sees not only the environmental issue but also the durability of the product. A product's durability—or, more typically, lack of durability—can translate into litigation for the supplier. That's something to be avoided, as recent class action litigation indicates. Suppliers also want to avoid the health problems associated with formaldehyde emission and volatile emissions from the plant.

Environmental and potential environmental issues are the byproducts of a rapidly changing manufacturing process, a process that increasingly relies on taking a small log apart into even smaller pieces and reconstructing it into wood products of beauty, durability, and enduring value. We have seen great progress in the development and refinement of adhesive systems and bonding technology. The question is, will progress be rapid enough to meet the needs of a growing world population? The coming decade will tell the tale.

Complexity and Technology: Dealing with People, Products, and Processes

10

Complexity and technological change will be the norm for the forest industry in the new century. The global marketplace will mirror the growing complexity of dealing with people, processes, and products. Interrelationships between political and technology issues—sometimes confusing, sometimes complicated—will challenge even the most adaptable and seasoned participant.

Charlie Bingham, an up-through-the-ranks Weyerhaeuser senior manager (now retired), probably said it best during an address in 1986:

> We [the forest industry] are in a period of major transition, in which the successful competitors will be those who are most aware, most innovative, and most responsive to change. The era of the bigs gobbling up the smalls is over. The era of the quick gobbling up the slow is upon us.

These prophetic words are true. The industry roster of the new century is quite different from the roster of 1986. Some "bigs" have disappeared forever; some are gone but don't know it yet; and some have adapted, changed, refocused their resources, and moved forward.

New firms have sprung up; some of these have withered and disappeared, others have prospered. The common thread among the survivors is a commitment to constant reassessment, learning from the process, and moving on. That's the goal of this chapter: reassess, learn, and move on.

Meeting the Challenge of Change

He's older now. I became reacquainted with this professor at the foot of Spencer's Butte on the southern edge of a long-ago timber town, Eugene, Oregon. The sparkle in Dr. Stuart U. Rich's eyes was still there as we discussed another time and my place as a student. As he related his daily regimen of walking to the top of the butte without stopping, I remembered fragments of a speech he gave nearly 30 years ago:

> People are expecting many more things from business than in the past. . . . Many companies have established excellent records over the years in meeting the

challenge of economic and technological change. There is no reason why they cannot similarly respond to the challenge of social change. (Rich 1971)

His 1971 speech advocated a more proactive approach to social change on the part of American business. Today, the forest industry deals not only with economic and technological change, but with the social changes brought on by the expectations of a growing and increasingly urbanized population. And this issue arises not just in America (although America does have the largest and most diverse forest products market on the globe). Similar expectations arise in other nations where citizens want a better life and a better environment. The end result is best described as "a frenzy of change . . . an epidemic of rapid change," as Dr. Jean Mater suggests in her book, *Reinventing the Forest Industry.*

How best can the forest industry leader respond to this "epidemic of rapid change"? The answers are many and varied; an important few follow.

- **Think Change.** The late Sir James Goldsmith did; he was quoted as saying, "If you just do the same as everybody else, you have to fail. It's inevitable. There's no escape . . . if you can see a bandwagon, it's too late to get on it." (Wansell 1987)

  Commit your organization and yourself to change; welcome change as a friend. Think about it, look for it, and encourage others to seek it. I'm reminded of a young lumber quality control supervisor I knew some years ago. He wanted to test new kiln schedules in a fir mill in western Oregon.

  "I don't see why we need to do that," remarked the dry kiln tender. "I've got thirty years on kilns. I know how they work."

  He was unwilling to try. Later, a coworker told the QC supervisor: "He probably doesn't have thirty years experience; he just been here. He has one year's experience, thirty times."

  The dry kiln tender was not alone in resisting change; a similar attitude permeated the entire business. The mill eventually closed—not for a lack of timber or markets, but for a lack of change.

- **Search Out the Knowledge of Others.** Takami Takahashi, a successful computer entrepreneur, wrote a decade ago: "Successful people surprise the world by doing things that ordinary logical people think are stupid. . . . If I listened to logical people, I would never have succeeded." (*Personal Computing* January 1990)

  The learning process comes not just from seeking knowledge of the forest industry and the technology that surrounds the business, but from getting inside the

Exhibit 10.1: Meeting the Challenge of Change

- Think change.
- Search out the knowledge of others.
- Welcome change and implement it.

minds of others, including those outside the business. The global industry and the expanding contacts with other cultures—and the thinking processes within those cultures—provide an unprecedented opportunity to gain knowledge from others. This knowledge can be translated into business success.

Some years ago, the need to avoid obsolescence was described to a group of prestigious Mark Reed scholars. "You need to have a lifelong personal program for reading and learning . . . the knowledge gained with the degree you are now earning will be obsolete in five to seven years."

After the speech a young scholar came up to the speaker with tears in her eyes.

"You really mean that? With all the work I'm going through to get my degree, you mean it will be obsolete?"

The speaker replied, "Knowledge is being gained so rapidly that what you are now learning will provide two things, a base from which to gain more knowledge, and thinking and analytical skills. Your degree work is well served if you learn both." (Baldwin 1987)

The forest industry participant needs to hear that same advice and take it to heart. Knowledge of the business, of people and how they think, and of the changing structure, technology, and products of the wood business—these are essential for survival in the changing competitive environment.

- **Welcome Change and Implement It.** Jack Welch, Chairman and CEO of General Electric, was quoted as saying: "If the pace of change outside exceeds the pace of change inside, the end is in sight."

  Mill closures frequently result from failures in planning and leadership, or a lack of commitment to the business. Too often, resistance to change is the culprit. The Heggenstaller lumber and pallet operation north of Munich, Germany, is an example of an adaptable—and durable—business, one that has managed change and maintained commitment for over 760 years. The mill complex, run by generations of Heggenstallers from 1855 onward, has manufactured lumber since A.D. 1230. The Linck Canter lumber line, installed during the mid- to late 1980s, is totally different from the earlier-generation sawmills on the site. This log processor and cutting mill represent a continuing commitment to change, and a commitment to the community and workers.

The Forest Products Enterprise: A Complex Adaptive System

Colin Crook, Senior Technology Officer for a large banking and finance group, observes complexity and technology from a business perspective "[with] the enterprise as a complex adaptive system—thriving in a world dominated by incredible technological change." (Crook 1995)

Crook views the enterprise as a vehicle bridging the gap between the exponential technological innovations and the chaotic market changes. Innovation and change are occurring in the business community at large, and the forest industry is no exception. The business will function as a dynamic bridge only by setting aside the knee-jerk operating methods still relied on by too many companies, and unleashing the entrepreneurial and competitive energies of the industry participants. For example, large forest product corporations go through so-called "profit improvement programs" in predictable cycles. These programs allow the older, more experienced personnel to retire with a golden handshake. Too often, this costs the company something more than the golden handshake: they lose irreplaceable experience and knowledge in dealing with turbulent and chaotic markets at all levels. It is not uncommon to keep the sales organization and jettison the marketing and R&D groups. Lower immediate costs soon give way to lost opportunities.

How can companies move beyond this "knee-jerk" way of doing business? They must learn to operate each day as if it were the bottom of the market, then restructure the organization to be flexible and adaptable. And in order to get ahead of events, rather than scrambling to react to them, the company must make investors and lenders part of the team.

Getting ahead of events is the key. The organization and its participants must assess, confirm, and adapt continually as events unfold. They must get ahead of the bandwagon.

A classic case of not getting ahead of the bandwagon—in fact, not even seeing the bandwagon—was played out in the Pacific Northwest in recent years. Industry participants watched as the timber harvest on federal forests plummeted. Most thought they needed a political solution. These people also thought the solution would be played out through lobbying in Washington, D.C. The forest industry failed to recognize that environmental issues were more important to the public than timber income and jobs.

As this dawned on participants, almost overnight they began flocking to forest industry trade association meetings. But at subsequent meetings, participants saw the ranks thinning, as additional mills ceased operation. In gab sessions before and after meetings, attendees began to focus on who wasn't there. The meetings became similar to a game of musical chairs; people began to wonder which chair and which player would be missing during the next round.

Those that had previously read the tea leaves began purchasing additional timberlands, efficiently processing small-diameter timber from private sources, updating their mills, and seeking new markets. Some timberland owners survived by shutting down their mills and selling logs. Spotting the bandwagon ahead of time, then taking action before the need arose, were their keys to survival. A complex adaptive organization requires timely and accurate competitive intelligence, thoughtfully processed on a continuing basis. The survivors made certain that was part of their operating strategy.

Forecasts, Models, and Operating Strategy

There is a wealth of forest industry information available through commercial and informal channels. The various Internet carriers provide instant information, most on

A measure of success: a centuries-old German community and its lumber operations and sustainable forest.

call at any time. A variety of consultants will conduct specific raw-material, product, or market studies for a trade group or individual. Sophisticated models, both simulation and linear programs, grind out answers as fast as the questions are generated.

Industry managers can readily acquire biweekly price reports, instant updates on earnings and earning potential for sectors and companies, and a variety of forecasts. However, these information sources assume that the industry is predictable within an acceptable level of variation; sometimes it is not.

Incoming information and the computer can only do so much. The dynamics of an ever-changing global situation constantly swing the cost and demand for raw material and wood products between regions and nations, seldom maintaining equilibrium for long.

A small change in market variables can cause sharp swings in prices and demand. Currency exchange relationships, monetary or fiscal policy, foreign exchange needs, weather events, consumer moods, transportation situations that impact the flow of goods between mill and customer—these are only a few of the thousands of variables that, singly or in aggregate, shape the operating environment for the forest industry firm. The successful operator makes money in this kind of environment by adhering to the following fundamental methods.

- Control your source of raw material, through ownership or predictable relationships. The cost has to be low enough to ensure at least breaking even during a down market period.

Exhibit 10.2: Keys to Making Money in a Chaotic and Turbulent Environment

- Control your raw material.
- Lock up your customer.
- Become a low-cost manufacturer.
- Ask "What if?" questions and search out the answers.
- Develop tactics that weather the downside and build the upside.

- Lock in your customers, either by providing a unique product at a fair price, or through supply agreements that provide predictability for both seller and buyer.
- Structure the manufacturing facilities so they are cost-competitive, with predictable performance.
- Ask "What if?" questions, eliciting a full range of answers, and assuming the worst case to ensure business survival.
- Identify specific operating-plan changes and business positioning that will protect the downside; then, based on the pessimistic scenario, formulate tactics to build the upside into the plan.

This strategy controls the three major variables—raw material, product sales, and mill performance—*then* focuses on forecasts and financial models.

Competing and Surviving in a Global Environment

Arie de Geus, former coordinator of group planning for Shell, described the fundamental question facing a business during a 1995 conference in London:

> What is the measure of corporate success? The dominant school of thinking on business administration measures success in purely economic terms: "maximization of profits, asset value or turn-over." A place in the FT 500 or Fortune 500 is the symbol of success. These criteria reduce the concept of a company to that of a money-making machine. Managing money-making machines is reassuring and comforting. It reduces the company to something which is rational, calculable, and controllable. The company becomes an Economic Company—the equivalent of Homo Economicus." (Crook 1995)

He continued by discussing an unpublished study sponsored by his company to answer this very question: "What is the measure of corporate success? How do you determine [it], and by what measure?"

The questions—and the answers—are further complicated by the global nature of the forest industry. We live in a world in which free trade is not necessarily fair, in which some governments are actively involved in providing raw material, employment, and

Exhibit 10.3: Business Killers

- Identity crisis
- Pride in past successes
- Following the herd
- Getting enamored with technology
- Cleverness in financing

markets to industry, and some are not, a world in which customs and thinking processes vary and the parties don't all play by the same rules.

One measure of corporate success is the economic model. Economic companies work with capital, labor, and other resources to optimize the return on capital employed—or a similar financially based indicator—to make a profit and maximize shareholder value. Overall, the company strives for efficiency; the managers seek maximum results with a minimum of resources. Strong internal controls make the task easier and more visible. The economic company grows because each quarter is expected to be better than the last, or the same quarter the previous year. These companies grow, acquire, merge, change their names, produce new and different products, and do all the tasks necessary to maximize shareholder value. But trouble can come in many forms.

- **Identity Crisis.** The growth of the company results in management's losing touch with the fundamentals of the business. Malaise sets in; senior management then tries the newest management fad to revive the business. Team management, re-engineering, quality circles, management by walking around, and management by objective are just some of the familiar fads of past years. Each has merit; however, few seem to last. The cycle begins again. Enthusiasm for each management model change diminishes with the passing years. Even the grand introductory dinners don't taste very good anymore.
- **Pride in Past Successes.** The past becomes the mantra of the future. "Never mind that conditions have changed, and that others are after the markets. It worked well in the past, and it should work well in the future. We just need to do more with less, faster, with more capital." The company takes incremental steps into the future, in the process overlooking the nimble newcomers who are not constrained by past successes.
- **Following the Herd.** Senior managers observe their peers and decide that if they are doing something, and spending a lot of money, it must be good for the company. So they build or acquire a lot of mills. The forest industry is a "herd" industry, building and producing the same products until it doesn't make sense anymore. Then participants wonder why they don't make money—and so do the shareholders.

- **Getting Enamored with Technology.** The thinking process usually begins with giving credence to the myth that the world is running out of trees. "If we can get the last little bit of recovery, we can make more with less." However, too often the mills become monuments to technical engineering skills, rather than money-making machines.

  Two North American companies are prime examples. Both sought to become wood-processing powerhouses. Company One built a mill, then built the next one with the same blueprints (with some revisions). They used high-tech equipment, but the return on investment (ROI) for each had to show up on the statement. Operating efficiency, low cost, hands-on management, and access to the customer were key.

  Company Two viewed the task differently. They designed each mill from scratch and filled them with the latest high-tech fads and fancies, many of which were invented by the company's R&D department. They developed a manufacturing sector with picture-perfect mills, each costing at least 40% more than the competitors'. Twenty years later, only a handful remain in operation. The company now sells trees to its competitors.

- **Cleverness in Financing.** Some companies just seem to hang on forever, particularly if they are public companies and tell a believable story. The forest industry has some classics. These companies seem to never make money, but exist by hope and cleverness. They issue stock, they buy back stock; they break into separate operating companies to create value, and they re-engineer, downsize, or merge to get the cycle going again. Occasionally they bring in new blood to jump-start the process. Sooner or later the end of the trail is reached for the manager, the company, or both.

These examples are both American, but these situations know no national boundaries. Asia, South America, Europe, and other North American nations all can provide prime examples. Arie de Geus is convinced that we must find a balance between short-term economic values and long-term survival.

> Shell research into companies of more than 100 years of age described these latter companies in very different words: "Successful corporate survivors are financially conservative with a staff which identifies with the company and a management which is tolerant and sensitive to the world in which they live." These companies behave as if they are living systems rather than mere instruments to produce goods and services. . . . Their assets are a means for living and "profits are like oxygen"; they need it to live, but it is not the purpose of life. (de Geus 1995)

The following are characteristics of long-lasting companies, based on the Shell study, which included a 700-year-old Swedish pulp and paper company;

- They seemed to be conservative in financing; "money in the pocket gives flexibility and speed." Cleverness in financing was specifically excluded.

- Each company was sensitive to their operating environment—very much a part of their world.
- The companies seemed to have a sense of cohesion and sense of identity. They had a clear definition of what they stood for as defined in a value statement
- Most made use of a decentralized structure, delegated authority, and had tolerance for issues.
- Capital was not considered the chief asset of the company—people were.
- They were adept at dealing with a changing environment, and had learned to make the company or organization respond to the changing environment
- They learned to develop new skills to exploit the changing environment in new ways.
- They changed the business portfolio several times over the life of the company.
- They did not insist on central control for new business activity.
- Tolerance was a measure of openness in the company.

The study went further:

> Any company aiming to live longer than some 50 years will experience fundamental environmental change over which it has no control. Ergo, it has no choice but to change its internal structures to restore the harmony with its environment. (de Geus 1995)

And it concluded:

> In the surviving company, personnel policies aim to create the space for innovative individuals, the mobility of the members and the means for social propagation which evolutionary biology indicates as essential for faster institutional learning. (de Geus 1995)

What then is the measure of corporate success? It's probably a sensible mix of the economic model and the survival model. Overall, it is a measure that fits the operating environment of the country and of the company's place in the world.

How do we deal with complexity and technology? Some of the answers were provided in this chapter; many more will become available as the forest industry proceeds into the 21st century. The essential task ahead: value the lessons of the past and actively shape the future.

Being Agile and Low-Cost: The Key to Profitability

11

Forest products and their nonwood substitutes now trade freely between nations and continents. As tariffs and other barriers decline, the importance of being an agile, low-cost producer takes on a new dimension. If the producer's manufacturing costs aren't right, nothing else matters. Medium Density Fiberboard (MDF) showcases this worldwide competition.

MDF now ranks with pulp, paper, construction lumber, and structural panels as a global commodity. With MDF now being produced on six of the world's seven continents, a plant in Indonesia can compete in the same markets as a South Korean plant or its Portuguese counterpart. The question has become, can the commodity producer provide the quality needed, in a timely fashion, at the just-right cost?

> It's the law of the jungle," says John Correnti, president and chief executive of Nucor Corp., the Charlotte, N.C., steel maker, which has reshaped its industry by building low-cost mini-mills. At the end of the day, it's the low-cost, high-quality producer who wins." (Aeppel 1997)

Although Correnti was talking about steel, the law of the jungle applies equally to MDF and other wood-based commodities. How can forest products be manufactured cost-competitively? What roles do the business leader and staff play? How does an organization create a work culture that achieves low costs and superior quality? The answers lie in the ability of a plant to manufacture a product cheap enough to offset the obvious trade inequities and currency relationships.

Bigger Isn't Necessarily Better

When big companies want to lower costs, they often think that building big plants is the answer. High technology is the norm; high production is the driver. Headquarters sends out the raw material program, the manufacturing schedule, and the order file. Managers overwork the letter R as they re-engineer, restructure, rationalize, redirect, or RIF (reduction in force) to lower costs. And then there are the D's—de-layering and downsizing. Management at the hands-on level hears of plans to "zero-base the budget," "organize into a virtual company," or "go to activity-based accounting" to improve results. But does all this work?

In their rush to build, many companies forget this rule of thumb: the bigger the plant, the greater its dependence on market share for a wider area. A huge, highly capitalized plant often finds itself producing into a glutted market, where supply exceeds demand. Low product prices are then the price of an order file. Pulp-and-paper has done that; lumber has done that; now it's the turn of the composite boards, such as MDF and OSB, to repeat history.

Leadership and Hardscrabble Values

The low-cost dynamics of the commodity markets require a business to adapt in order to survive. The transformation process of assessing, evaluating, and changing must become a way of life. Enita Nordeck had it right when she and three colleagues founded Unity Forest Products one December day in 1987. Unity isn't a large-scale manufacturing plant; it doesn't have international competition. What it does have is regional competitors—plenty of them. Like its competitors, Unity remanufactures industrial lumber into a variety of products. But Unity's approach to the business—an approach that can be successfully used by its far larger commodity-based cousins—sets it apart.

> Unity's steady success hinges on the hardscrabble values that created it. Its 38 employees work as if they're on fire. The company makes sure that profit centers are indeed profitable. It combines speed, fair pricing, a keen knowledge of its customers' needs, and an intense symbiosis between sales and manufacturing with a near-fanatical attention to financial detail. In short, Nordeck runs the company with tactics that are nothing less than revolutionary in the hidebound, unhurried lumber business. (Davis 1988)

Unity's "hardscrabble" ways give a new dimension to that term as it's usually defined: "yielding or gaining a meager living by great labor" (*Webster's New Collegiate Dictionary,* 523). Unity's definition has a far wider scope. It means selling a product before the raw material is purchased or the product made. It also means a near-fanatical attention to financial detail and the collecting of monies owed. It means timely communication and participation with employees so each feels a part of what

Exhibit 11.1: The Inventory Ethics

- Keep the raw material fresh.
- Stay in the market using "just in time" (JIT) purchasing procedures.
- Maintain physical control of the inventory.
- Avoid old or obsolete inventory.

is going on. It means continually measuring, validating, and adapting as markets and customers' needs change.

Inventory management is an integral part of the hardscrabble culture. This is the inventory ethic:

- **Keep a fresh supply of raw material coming in.** This provides rapid feedback to the buyer and the supplier as the lumber is converted into product. This practice also allows the customer to constantly process fresh stock and minimize the tying up of cash.
- **Follow "just in time" (JIT) inventory procedures.** This keeps the supplier and the customer always on the market, and removes speculation from the buying and selling process.
- **Use inventory as a tool to discipline production and cash.** Taking a physical inventory at least weekly will help maintain this control.
- **Avoid jags (partial units) and black (weathered) lumber like the plague.** Follow this rule, and you won't waste cash on carrying costs, or labor on working around the unwanted lumber.

A hardscrabble culture makes the most efficient use of resources, rationing each to the best use. It promotes customer reliance and loyalty; the supplier's attention to detail makes quality and timely shipment an operating norm. The hardscrabble ethic looks for the cheap way to operate a business, but cheap doesn't mean greedy. Staff are paid well—often more than their better-capitalized competitors. Compensation includes both a fixed amount and incentives. The incentives, broad-based and easily measurable, usually relate to profitability of the business rather than statistical events such as production goals. Incentives tend to be short-term, with prompt, visible feedback measured in statistics and dollars. Accounting plays a crucial role in the bird-dogging and scorekeeping. This "playing to the statement" is a way of life in a hardscrabble competitive culture—as I learned many years ago.

The Accountant and Cost Accountability

Forest Gilchrist was the veteran controller for Simpson's collection of three softwood plywood mills in western Washington. I was the location manager, recently appointed to direct the Shelton mill operation through a multimillion-dollar modernization. During the winter of 1974 the mills were operating on a short schedule because of a poor order intake and low prices. A week-long curtailment became necessary. The intervening quarter-century hasn't aged the lessons I learned then.

During the downtime, I organized a skeleton crew to fix up, paint up, and clean up—all the things that never seem to get done during a normal production schedule. Each project was either a must, need, want, or nice to do.

Forest left the office and toured the mill, then came back into the office.

Converting logs in a well-designed mill is a key to profitability for some.

"Dick, all you're doing is adding to our losses; I didn't see a thing that will make money for us."

"But you don't understand, Forest, this may be the only time we get to do these projects!"

"But you don't understand, Dick—if we aren't making any money, then we shouldn't be spending any, except for the must-do projects that can't be done on scheduled production downtime such as weekends and holidays. You have got to play to the statement."

My first inclination was to verbally demonstrate the "rightness" of my position, but I paused to think. *What is more important? The financial performance of the business—or whether my direction to the crew was right or wrong?* Once I dropped my defenses—and overlooked the blunt directness of Forest's remarks—I saw several things clearly:

Exhibit 11.2: Playing to the Statement

- Hear *what* people say, not *how* they say it.
- Listen to your people.
- Look for new perspectives on the business.
- Do only the essentials.
- Keep track of costs, performance, and profits.

- **It's not *how* people communicate, it's *what* they communicate.** Don't allow yourself to be turned off by the way information is presented.
- **Staff people bring a new view to a problem or opportunity.** The leader needs to listen openly, inviting staff to express their thoughts.
- **When a mill goes down, every activity except the most essential should cease.** Musts and wants should be scheduled for regular down periods, when the statement can better support those activities.
- **An operating mill should be kept as lean as if every day were the bottom of the market.** This may mean sacrificing some upside, but the cost avoidance will yield big benefits during a normal market or periodic downturn. The mill will be more profitable over the longer term.
- **Keep track of costs, performance, and profits.** A monthly profit-and-loss statement is insufficient as an aid to timely decision-making. At a minimum, weekly statements should be available as scorecards and decision aids, supported by accurate weekly physical inventories.

Both the leader/manager and staff at all levels must have a deep and abiding respect for cost avoidance. One southern manager nearly wrecked a family-owned chain of small discount stores some years ago, until he remembered one of his dad's homilies: "Never forget the finality of costs, nor the unpredictability of benefits." (Stern 1993)

One mill did that, and cost its customers a supplier and 450 jobs in the community. Because of a disconnect between costs and purchasing decisions, the mill wasn't charged for supplies until they were retrieved from the storeroom, rather than when they were received. The managers' "Don't spend" admonition had reached the crew, but not the purchasing staff. Costs appeared to go down in the mill as the market declined, but the mill ran out of money with a buildup of nearly $750 thousand of unused supplies in the storeroom.

A "really need to do" project must play to the statement. Such a project either makes product or protects the people resource. It is a visible contribution to the business.

Inventiveness: The Hardscrabble Way

Inventiveness is a hardscrabble cultural norm, an outgrowth of the envisioning process. First, management describes an agile, low-cost business with a clear, compelling vision, then plans a strategy to meet specific, measurable goals. Inventiveness soon follows; benchmarking against the "best of the breed" accelerates the process.

Benchmarking helps you figure out where you want to be and how big a gap you must bridge to get there. The best-of-the-breed analysis compares your operation against the best in your company, your industry, and other industries. The process establishes a competitive framework of comparison: Who are they? What are the drivers? How do we catch up or keep ahead? How do we maintain the competitive pressure?

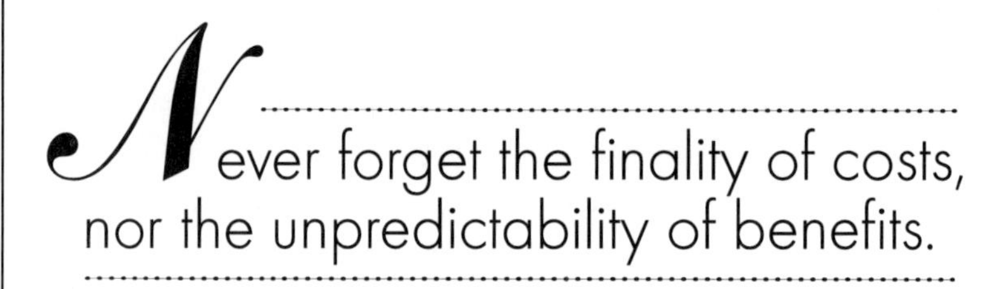

Forbes Magazine, November 8, 1993, p. 212

Exhibit 11.3

Technology is not an end unto itself; it enables change to upgrade the mill's competitive position. Frequently, a mill with an agile and competitive crew will outperform a high-technology mill costing five times as much, or more.

> In most plants, the boss stays busy managing materials, methods and manpower alone. But when the operation is committed to constant productivity gains, the plant manager shoulders a further load: as a curator of human creativity. "You can only work and sweat so much; it's finite . . . That leaves you with finding better ways to do things. To my knowledge, that's not finite at all." (Petzinger 1997)

In an inventiveness environment, a diversity of people work in teams, which in turn promotes a diversity of ideas, in two ways. Type A takes the old building blocks and puts them together in a new way. Type B invents new building blocks. Identifying Type As and Type Bs is the task of leaders and crew. Teamwork is institutionalized, but not the teams themselves. Diversity creates tension; so do ideas. The leader's presence on the floor is often the catalyst for bringing out new ideas, sharing information (which sparks additional ideas), and identifying barriers to change. Mark Schmidt, the founding plant manager of Dana's Stockton truck chassis plant in Stockton, California, was a leader whose team selection and leadership skills make the difference.

> Convinced that a diversity of people promotes a diversity of ideas, he [Dana's Mark Schmink] wound up with 19 different nationalities in a workforce totaling less than 300. Diversity, of course, often causes tensions, so Mr. Schmink became a constantly visible presence in his blue overalls, demanding to know when anyone felt slighted, offended or put out. (Petzinger 1997)

Teams generate ideas and more ideas. One Dana employee initiated 180 ideas during 1996—an astonishing example of what an inventiveness culture can produce. Because most forest industry operations are located in rural communities, their leaders will be hard-pressed to accomplish the Dana feat of 19 different nationalities. However,

Type A takes the old building blocks and puts them together in a new way. Type B invents new building blocks. Identifying Type As and Type Bs is the task of leaders and crew.

Exhibit 11.4

managers can still accomplish much by seeking diversity in jobs, backgrounds, and interests when assembling a team.

An inventiveness culture is apparent in the forest industry's top performers. Their leaders believe in people, value new ideas, and recognize the business necessity of change and innovation. This is especially apparent in the fast-growing Engineered Wood Products (EWP) sector of the business.

Teams that bring together sales people, engineers, architects, foresters, mill people, staff, and customers in work groups bring new excitement to the business. The designing and championing of products has changed the character of the building industry. The construction materials at the beginning of the 21st century are a world away from the materials of just 10 years earlier. I-joists are now becoming a job site staple; structural panels and LVL are becoming commonplace.

Thinking and the Road Ahead

Innovative thinking was a prized attribute in early logging and mill operations. Everybody did it; it lightened the load, it sped up the process, it improved the product, it helped an operator survive and get ahead. Then the industry went through a long era of standardization and mechanization. Too often, innovative thinking was left to the salaried staff, the engineer, and the people in the laboratory. With the current forest industry restructuring, there has been an emergence of new raw materials, new technology, and new products. Adaptation, innovation, and change have become prerequisites to survival. The high costs of labor, equipment, and raw materials and the oversupply of commodity products have resurrected old-fashioned innovative thinking as the ticket to survival and profitability in the 21st century.

Section Four:

Marketing and Sales: Meeting Global Demand

12 Population and Consumption Trends in North America and the World

The forest products industry is one of the world's leading producers of goods and services: about 3% of the world's gross economic output is generated by this sector. The output of tangible goods and services is huge by any measure. However, traditional markets, products, and customer relationships are changing in response to population growth, consumption trends, and wood availability.

Trade agreements—such as in the North American Free Trade Act (NAFTA) and the World Trade Organization (WTO)—and the realization that the earth is a relatively small planet are causing the world's political and business leaders, nonprofit groups, and individuals to rethink the value of the tree. The predicable outcome: changing roles for the public, the manufacturer, and the forest resource manager. An analysis of these changing roles will provide an in-depth understanding of forest use and consumption trends.

The emerging trends are categorized as one of five so-called "Re's"—Redefining, Reforesting, Redistribution, Recycling, and Re-engineering. Forest industry players must understand the role of each Re in order to identify the emerging consumption trends.

The Five Re's

If we think in traditional ways, we conceive traditional ideas and solutions. Our essential task ahead is to disregard or, at the least, re-evaluate the conventional wisdom of the past, and move on as proactive seekers.

Traditional wisdom would have you believe that there is an impending fiber shortage. Gifford Pinchot and Teddy Roosevelt said that nearly a century ago. Federal land set-asides, which later became our national forests, resulted from this belief. Today, belief in the pending destruction of forests and species is used to prevent harvesting. Many would have you believe that tree harvesting will destroy a forest; that does not have to be the case.

In fact, there is no impending shortage of wood fiber, so long as we manage tree growing, tree use, and wood reuse carefully. The world population is expected to grow from 4.5 billion people to 7.4 billion in the 40 years between 1980 and 2020, an increase of about 166%. In the same time period, roundwood harvests are expected to climb

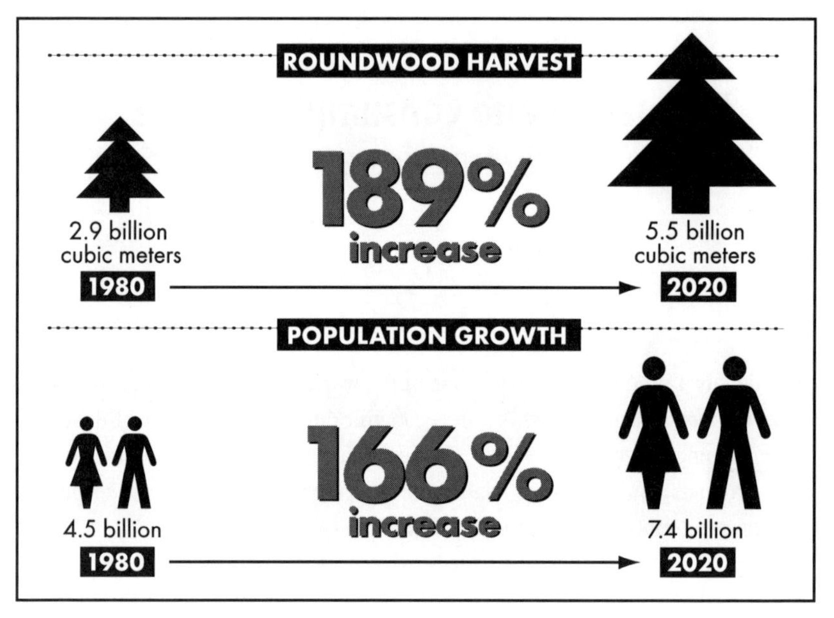

Exhibit 12.1: Comparison of Population and Roundwood Harvest Growth Rates

189% on a sustainable basis (see Exhibit 12.1) And the demand for wood fiber is increasingly being met by use and reuse of other organic and inorganic materials.

Global roundwood production, 2.9 billion cubic meters in 1980, is expected to climb to 5.5 billion by 2020. Although it's true that some of the emerging substitutes for traditional roundwood aren't situated conveniently to the consumer, may not be grown in traditional ways, and may not look like traditional wood, they are all wood fiber. And the industry will figure out how to transport and utilize these new sources.

But we won't do so without challenging traditional norms. The conventional definition of forest ownership classifies the forest as either state-owned or privately owned, with the owners and successors becoming the beneficiaries of the forest. In the emerging economy of the future this will not necessarily be so. The "tree huggers versus the tree cutters" paradigm will shift to another view of the forest and ownership.

Central to the emerging paradigm are the beliefs that man is a guest on the planet and that each generation must consider the one to follow. The forest benefits now recognized are more numerous and far-reaching than ever previously imagined. Increasingly, forest benefits are described in terms such as carbon sequestering, biodiversity, ecosystem management, sustainability, climate health, and cumulative effects. Forest owners can now be categorized as owner/stewards for everybody else. The role conflict between the owner/stewards and the "everybody else" category presents a challenge as we sort out the issues of ownership and public need for a wider range of forest benefits on a global scale.

Around the world, neighbors and trade partners seek to resolve the issue of who gets what, and at what cost. The tree-growing steward may receive income not only from wood products but from the other products of the forest, such as wildlife, clean air, and clean water.

Let's take a look at those five Re's of forest products consumption, and how each may define the consumption patterns of the 21st century.

Redefining the Raw Material

The world population is expected to increase 166 percent between 1980 and 2020. In the same time frame, printing and writing paper production and consumption—a good indicator of developing wood markets—is expected to increase 442%, while wood-based-panel consumption can be expected to grow 207% worldwide. To meet this demand, it will be necessary to redefine the raw material.

The traditional notion that lumber is made from a sawmill-type log, and that solid wood will continue to dominate rough and finished carpentry, is passing into history—along with the concept of long tree rotations for sawtimber, and the need to rely on a particular favored species.

Most wood fiber—whether large or small, tall or short, hardwood or softwood—will be fair game for the raw-material specialist. Product manufacturing processes are available to manufacture the increasing variety into almost any product. The major consideration is cost: cost of transportation to the mill, of manufacturing, and of transportation to the customer.

Technical innovations will produce such required attributes as strength, beauty, and utility. Solid wood furniture, traditionally constructed with an exotic species, can exhibit the same beauty and utility with alternate technology. Furniture cuttings that develop from plantation-grown wood form the inner core. The exotic species, obtained from a sustainable source and peeled into a 1/42-inch (0.6-mm) thickness, is laminated to the core to form the outer surface. This alternative construction provides the beauty of the original design while conserving more valuable wood.

Comparable beauty and utility can also be obtained with the clever use of stains or nonwood overlays, each over a composite core. The conservation imperative of a shrinking volume of old-growth trees, combined with technical innovation, provides the sought-after products of the past in a new form. The same innovation skills are providing newer, more economical forest products. The lowly 2 × 10 (51 × 254 mm) is a case in point.

The I-joist, which features a parallel veneer laminated flange and OSB webbing, is a competitor to solid wide lumber. I-joists are now being manufactured cost-competitively as a substitute for 2 × 10s as well as wider, thicker beams. It is not surprising that the traditional price premium for 2 × 10s versus narrow lumber is disappearing. Wide lumber often sells for about the same price as 2 × 4s (51 × 102 mm) on the MBM, or million-cubic-meter, basis. The redefined raw materials are rapidly gaining acceptance from both the manufacturer and the consumer.

Reforesting the Land

Redefining the raw material is closely related to the second Re, reforesting the land.

> The industry-wide drive for standardization and consistency is moving down the value chain from final consumer products through to the forest. . . . Eventually, this trend will lead to more investment in processing assets that can guarantee consistency, and a movement toward either tree plantations or homogenization during primary and secondary processing. (*Business of Sustainable Forestry*, 2.2)

Both process homogenization and reliance on tree plantations are growing more common. The two, acting in harmony, result in highly automated manufacturing facilities—facilities that process a uniform raw material that is further manufactured into value-added products. Two reforestation models predominate: the natural forest model and the plantation model.

The natural forest model is more prevalent in the developed nations; the U.S. Forest Service has adopted this model. Under this model, the resource manager follows the principles of ecosystem management, managing the natural forest for a host of values, including timber production, wildlife habitat, clean air and water, and scenic beauty. Wood production is just one of the many forest uses. The Detroit District of the Willamette National Forest is a case in point.

A sustainable annual harvest of 117 million BM per year has been curtailed to 9 million BM. The road system, built to protect the timber asset and to provide access for harvesting, is now being slowly abandoned as these forests are shifted to ecosystem management. As the debate continues among a variety of constituencies—including the Forest Service, environmental groups, the forest industry and other interested parties—and because of continuous litigation or appeals by various interest groups, the realigned values are considered to be more important than locally produced lumber and jobs. Lumber can be gotten from Canada; the jobs can be shifted to other industries. Abandoned homes in the adjacent timber-based towns and accelerated harvesting in a neighboring nation are considered to be an acceptable trade-off by the growing urban population that has obtained greater say in how the forest is managed.

Traditional timber-growing and agricultural lands in other temperate areas in the Northern Hemisphere and throughout the Southern Hemisphere are being converted to single-species plantations. This forest model is viewed as the most efficient way to produce wood fiber and reduce atmospheric carbon emissions to offset the pollution created by the developing and developed nations. The essential components of this model are advanced genetic programs, maximum yield management through the use of fertilizers and other soil amendments, and rigorous integration with downstream processing facilities such as chip mills, lumber operations, and board plants.

Single-species plantation management has become the dominant form for growing softwood in the 10 southern states that stretch from Arkansas to Virginia. However, the

A modern German log processor near Frankfurt.

single-species 28-year rotation is being dwarfed by the growth and harvest rotations in Central America and farther south in the Southern Hemisphere. There, vast single-species plantations grow both hardwoods and softwoods.

It is typical to have an 8- to 10-year rotation for gmelina (*Gmelina arborea*) at the Ston Forestal plantation operation near Gulfito, Costa Rica. On converted savannah lands in Brazil between the Amazon River and French Guiana, Caribbean pine (*Pinus caribaea*) is presently clear-cut at the end of a 9-year pulpwood rotation. The pine plantation lands in southern Brazil are also productive; the Brazilian sustainable harvest may exceed 466.7 million cubic meters as the harvest of the rainforest winds down and additional plantation forests come of age (see Exhibit 12.2).

The growing conditions of this South American nation could enable it at some point to surpass the United States in roundwood production. Australia, New Zealand, and South Africa have also discovered the plantation model. Although these nations lack sufficient land to provide huge roundwood quantities to world markets, they more than make do with what they have. The three nations combined will produce a 10% to 13% share of the world's developing roundwood supply for the foreseeable future, mostly from plantations.

The two forest management models will provide the roundwood fiber needed for the future. As additional forestlands are moved into the ecosystem model, more lands must be converted to the plantation model as a source of supply. However, a huge potential gain in useable wood supply will also come as a result of the third Re: Redistributing the roundwood to higher and better uses.

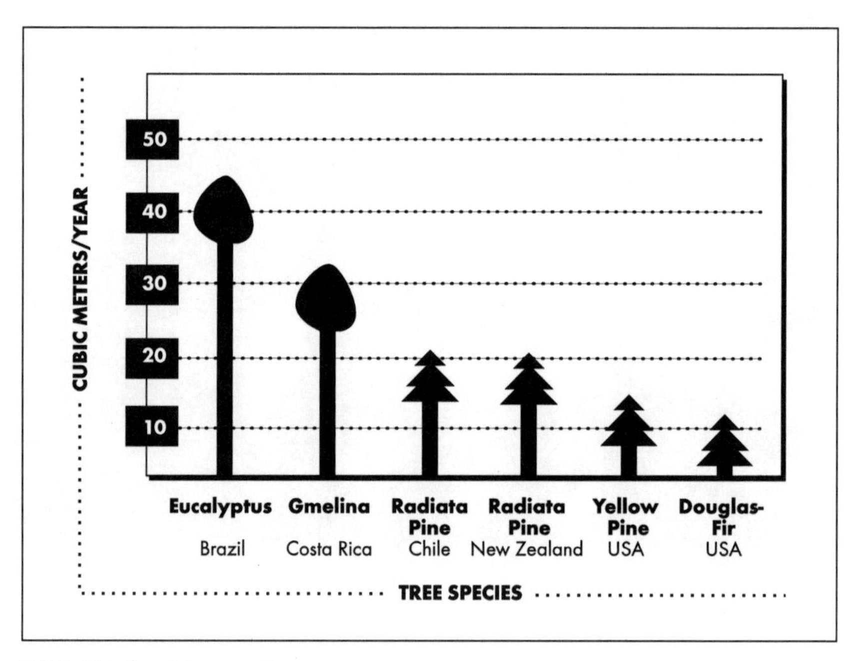

Exhibit 12.2: Growth Rates per Hectare

Redistribution to a Higher and Better Use

Is the industry using the roundwood to its highest use? And using it efficiently? For over half of the harvest, the correct answer is "Probably not." The world's political leaders and business people should be pondering these two questions. The facts are sobering.

More than half of the world's roundwood harvest is used as a cheap energy source, for fuelwood and charcoal. Additional volumes are lost as the land is cleared for subsistence agriculture. The 1995 global production of roundwood was reported at 3,411 million cubic meters, and in that year about 1,923 million cubic meters, or 56.3% of the total, was consumed as fuelwood and charcoal. The 1995 fuelwood and charcoal percentage is higher than any recorded for the previous 5-year intervals of 1980 (50.5%), 1985 (51.7%), and 1990 (50.7%), as reported by the Food and Agriculture Organization (FAO) of the United Nations.

A further review of the 1995 FAO statistics yields the following information:

- Six nations—Brazil, China, India, Indonesia, Russia, and the United States—produce about one-half of the world's production of roundwood.
- The fuelwood and charcoal consumption of these six nations—55.9% of total roundwood production—nearly matches the worldwide figure of 56.3%.
- Two nations, Russia and the United States, consume 23.5% and 18.8% of their roundwood, respectively, as fuelwood and charcoal.

Clear-cutting a short rotation of plantation in the Southern Hemisphere.

- One underdeveloped nation, India, consumes 91.4% of its roundwood as fuelwood and charcoal. Indonesia (81.3%), Brazil (70.3%), and China (67.9%) are also major consumers of fuelwood and charcoal.

These last four nations together contain about 43.5% of the world's population. The per capita consumption of wood products will increase as their economies develop, with a growing drain on the world's resources to satisfy that demand. Most of their wood products will come from global markets. This is particularly true for the nations, such as China, that lack sufficient growing stock of their own.

Those nations with abundant wood supplies have a dual opportunity; they can reduce fuelwood and charcoal consumption, and also upgrade their wood-products conversion facilities, thus obtaining more products from less fiber.

A Central American mill provides an example of what not to do. Its Scragg primary log breakdown system has a saw kerf of about 1/4 inch (6.4 mm). This mill saws pallet stock out of an 8-inch (2.1-cm) diameter log. The actual lumber yield is only about a third of the log; the remaining two-thirds is conveyed onto an adjacent sawdust pile, into indefinite storage.

Due to lack of knowledge and capital, too many other new mills in the region, and in other nations, operate the same way. Sometimes this lack of knowledge even extends to the technical advisors provided by the government. Lack of knowledge, lack of capital, and, in some cases, a negative bias against newer, more yield-efficient equipment is literally robbing the forest of its trees.

The world's forests are in the spotlight, but that spotlight has been placed on the developed nations that have the highest conversion efficiency, with comparatively little attention focused on the redistribution of the roundwood harvest to its highest and best or most efficient conversion methods. The industry must turn its attention to this task.

Recycling the Resource

The beauty of wood is more than skin deep; its utility extends to the innermost fiber, which stores carbon indefinitely. Wood is a carbon sink; the tree takes CO_2 as a gas and converts it into fiber. However, the process extends an additional step—one that many otherwise-knowledgeable people find difficult to understand and accept.

When a forest matures, the growing cycle begins to decline. When, in time, new growth ceases, the forest shares an attribute with wood furniture: it becomes a carbon sink of beauty. An abundant and vigorously growing forest plus the wide and growing use of wood products represent the optimum conditions for both the wood products consumer and the consumer of other forest values. Recycling has a valuable role to play.

The theory of recycling goes something like this: the more you reuse the fiber, the less you need from the forest. By using less from the forest, we also have less need to disturb the standing wood as a carbon sink. Some believe that recycling paper and paperboard is the total answer to protecting the forest. Unfortunately, this is a practical impossibility. After about four or five times through the pulping process, fiber will turn into mush and be flushed down the drain. Even with the most ambitious recycling program, we will always need to add virgin fiber to the pulping process. But this should never discourage us from recycling.

The paper and paperboard industry has demonstrated effective recycling. The fiber furnish used to manufacture the world's paper and paperboard totaled 293 million metric tons in 1995. Of this, 40% was recycled raw material (see Exhibit 12.3). Recovered paper has become an important source of raw material for the paper industry.

The recovery of paper and paperboard in municipal solid waste in the United States is another example of effective recycling (see Exhibit 12.4). The recovery rate more than doubled in a quarter century, from 16.7% in 1970 to 40% in 1995. And this does not include the private recovery efforts of others. The 1995 U.S. recovery of paper and paperboard represented nearly one-third of the world's total. North America and western Europe are the serious recycling regions in the West, with Japan, Korea, and China the heavy hitters in the East. Nation by nation, the trend will edge upward as the developing nations use more paper in both absolute and per capita terms.

China's paper industry is characterized by a large number of relatively low-tonnage paper mills. These mills, scattered throughout the major population centers, are strategically located to collect and transport the recycled material to the pulp mill. The economic benefit is huge; every ton of reused fiber means one less ton imported.

The fiber recycling effort extends beyond paper and paperboard. Reconstituted-board producers are beginning to tap into this cheap source of wood fiber for particleboard, medium-density fiberboard (MDF), and other fiber-based boards. The efforts

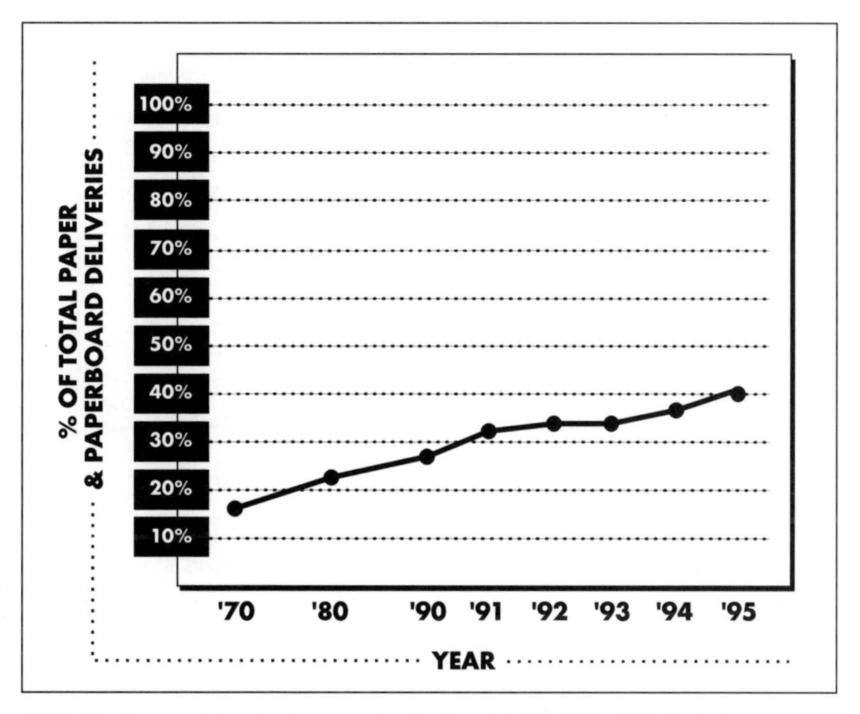

Exhibit 12.3: Paper and Paperboard Recovered by Year from Municipal Solid Waste Receipts

being made to recover and recycle pallets, wood debris, and other, like materials near major population centers have given rise to the term "urban forest." Some firms even specialize in tearing down aged wooden structures to recycle the old-growth beams, timbers, and other lumber products. The resulting lumber is often worth more than the standing building.

Some recycling efforts, such as the salvaging of old lumber, provide a relatively minor source of fiber. However, the trend worldwide is to reuse the fiber resource as a less costly way to obtain needed raw materials. The global benefit in aggregate is compelling: more products from fewer trees.

Re-engineering Wood Products

Redefining the raw material, reforesting the land, redistributing the harvest to a higher and better use, and recycling the wood-based product are major ways to meet present and future demand. As a result, the emerging raw material is materially different from that of the past. The forest products industry will find new opportunities in matching the available raw material to consumer expectations for product appearance and utility. The Laminated Veneer Lumber (LVL) industry has found phenomenal success in this area.

Exhibit 12.4: Paper and Paperboard Recovered by Year from Municipal Solid Waste Receipts

Year	Percent of total paper and paperboard deliveries
1970	16.7
1980	21.8
1985	21.3
1990	27.8
1991	31.7
1992	33.0
1993	32.9
1994	36.5
1995	40.0

Source: *Statistical Abstract of the United States 1997*, no. 385, p. 237.

North American production of LVL increased 20% in 1997, continuing a steady growth trend. This industry segment, a veneer-and-adhesive assembly used as a lumber substitute for beams, rafters, and scaffold planking and in the manufacture of I-joists, is expected to at least double in production and consumption every 5 years into the 21st century. Essentially, the process uses small logs peeled into veneer, plus adhesives, and new technology to manufacture a large log product from a small-log resource. The new product often performs better than the old, and is often cheaper.

LVL is one of the more visible of the product innovations that benefit from advances in wood, adhesive, and related technologies. Designers can now produce board and lumber products from increasingly smaller wood segments, which are in turn engineered into larger components with wood and adhesive technology. The flexibility of wood—which allows the design of infinite sizes, shapes, and features—provides the forester, the geneticist, the wood scientist, the wood engineer, and the architect with limitless opportunities to design new products that replace and improve upon the old.

The five Re's reflect and respond to the population and consumption trends of the present and future. Those in the forest industry who learn their Re's will understand what the consumer needs and expects—in both wood products and other forest values—and will be well prepared to successfully meet those needs in the 21st-century marketplace.

13 New Millennium Marketing

Lumber, panels, and other forest products sold in bulk have traditionally been considered commodities. The company produced a product, took it to market, and negotiated the best deal it could get. Normal product life cycle was measured in decades or lifetimes. However, different raw materials, new technology, innovative processes, and changing consumer preferences have fostered the development of new or redesigned products at an unprecedented pace. Industry participants must understand the needs of this changing marketplace and compare them to the traditional marketing model; only then can they revamp their marketing approach and realize the potential benefits brought by these changes.

The Changing Marketing Model

Historically, the forest products producer made a product—probably much like its competitors' products—and took the product to market, either through wholesalers and distributors, or directly to the customer. The customer shopped an order until the best price and delivery date were assured. During the peak demand portion of a market cycle, the mill had the upper hand in establishing price and delivery dates; low demand would rapidly reverse the situation, giving the customer the advantage.

Most, if not all, of a producer's marketing effort is provided by industry associations, which are organized by product grouping. APA–The Engineered Wood Association, the National Wood Window and Door Association, Hardwood Manufacturers Association, and Southern Forest Products Association are a few of the many U.S. trade groups that promote products on behalf of their membership. In other nations, similar trade groups or government agencies perform the same function.

Trade promotion includes national media spots, international promotion events, meetings with code officials and specifiers, and the production and distribution of literature that describes the product and its use. Company advertising and product promotion sell products through trade publications, point-of-sale displays, word-of-mouth promotion, and a heavy emphasis on price. Although computers and telecommunication advances have refined the marketing effort, making it more timely and efficient, the process has otherwise changed little over the years.

Dr. Sam Sherrill, executive editor of C. C. Crow Publications, had this response when asked what he would do to improve the traditional marketing paradigm:

> I'd be in their [the customer's] face . . . promoting and delivering on a wider mix of products, immediate service, quality that is better than most, and price relationships which allow me to be competitive. I would target markets in which I can do well and have a competitive edge on others. (Sherrill 1999)

Dr. Sam's response is right on. Many industry participants are following his advice and obtaining improved results. A consistently profitable panel plant in Mississippi does it this way: Give the customer what they want, when they want it, and in the amount they want, even if that is less than the standard bundle. If the customer buys in bulk, they may find it cheaper to go somewhere else, but if they need an order fast or need it to finish a job, this plant makes it and ships it within hours if necessary. This plant gets the business—and it is good business.

The Customer Defines the Need

These marketing refinements focus on the customer, and each refinement is an improvement over past services—particularly when the mill is competing against offshore suppliers who require high volumes and long lead times.

An important element of the new marketing strategy is letting the customer define the product. This is a distinct shift from the more traditional "produce the product and find the customer" mode of operation. Ironically, an incident that took place in an earlier era demonstrates the marketing tack that is consistently needed in the new century. Ed Westman, manger of Washington Veneer Company in the 1920s, reported:

> . . . that his personal visit during 1927 at the Haskelite Corporation factory in Grand Rapids had resulted in an order for a million feet of 13-ply 1 1/8 inch plywood for floors in busses and streetcars. The superintendent whom Westman believed responsible for the order told Ed he was the first plywood man to visit their factory to learn their problems, instead of "telling them what to do." (Plywood Pioneers Association 1971)

Exhibit 13.1: The Changing Marketing Model

- Customers define the need.
- Producers educate and respond.
- Quality is the report card.
- Informal and formal partnerships are essential.
- Communication and coordination are critical factors.

Trus Joist MacMillan represents a new generation of forest product companies that incorporate the customer needs typified by the Westman anecdote into their sales programs for engineered wood products. Trus Joist manufactures a huge variety of products, each with a specialized use defined by either the customer, the specifier, or the architect. Engineers design the product and specify what is needed for each job, based on customer input. The engineer is the right arm of the salesperson, who has ready access to software and engineering assistance. In this way, manufacturing can provide just what the customer needs—whether that be a floor system, roof trusses, or other wood-based structural components.

Other manufacturers are following Trus Joist's lead. Laminated veneer lumber (LVL), a relatively new and innovative product, is providing the spark to move away from the traditional marketing model. Sales benefits are impressive. Production increased 20% in 1997 and was forecast to rise 25% in 1998.

The hardwood plywood industry is an old industry seeking new answers in marketing.

> The hardwood plywood industry is very competitive. Comprising the industry are many manufacturers with relatively homogeneous products. When placing an order, buyers generally have the option to choose from several suppliers and usually receive several price quotes before placing and order. . . . For a firm to remain competitive in such an environment, the implementation of a successful marketing strategy becomes essential. The buyer's decision to purchase from a certain supplier is based on the buyer's perception of which supplier offers the optimal mix of product and service attributes. (Forbes, Peck, and Altman 1996)

However, listening to the customer is just part of the answer. Educating the customer is also important.

Educating the Customer

Don Davis, Sr., for many years a prominent Chicago jobber and wholesaler, provided a pioneering example of customer education:

> In the summer of 1928, Don [Davis Sr.] had seen some compressed paper wallboard in a vacation cottage in Wisconsin. The wall panels had shrunk and warped badly, sagging and tearing away from the nails. Don was sure plywood would do a much superior job and immediately wrote to Ed Westman . . . suggesting the manufacture of a wallboard grade of fir plywood 1/4" thick, 48" × 96". Westman pounced on the enterprising suggestion and in a short time . . . started receiving orders for 1/4" wallboard in carload lots. (Plywood Pioneers Association 1971)

The duo of Davis and Westman pioneered the notion that the best sale came from answering a need and educating the customer, rather than just peddling a product.

Educating the customer is a vital element of the new paradigm. As new products are introduced and old products disappear, the customer doesn't always know what is available. The supplier who does the best job of educating also does the best job of selling. Steve Boyd, president of Manufacturers Reserve Supply of Irvington, N.J., put it this way:

> A key ingredient of the wholesale distributor asset base is product knowledge. Through close relationships with their manufacturer partners, wholesalers work to develop new products, introduce products and then represent those products in the market. This entire process shortens the channel dramatically not just in the flow of products, but also the flow of valuable product knowledge. (Johnson 1998)

The manufacturer, the wholesaler, the distributor, and the retailer are all becoming links in a chain of educational activities that connect the producer and user. Educating the consumer grows in importance as nontraditional raw materials replace the historic mainstays of forest products.

Denny Ryan of Home Depot recently described what needs to be done when introducing a new product: "The seller has to know the product, and he needs to know the product well enough to point out the features which may detract if not understood by the customer."

Ryan's example took place on a visit to a potential suppler in Costa Rica. They were discussing teak flooring samples produced from 7-year-old plantation trees. The face appearance was attractive; the pith streak on the backside was not. But when attention was drawn to both the natural beauty of the product and the favorable environmental consequences of using plantation wood rather than rain-forest timber, the pith streak seemed inconsequential. Ideally, the seller will describe the benefits before the buyer notices the minor flaws.

The method of education described by Denny Ryan is important to the process. A recent survey of architects and engineers stated that the most influential methods for obtaining product information are:

- manuals/data files relevant to the product
- well-prepared reading materials
- word of mouth
- physical examples

Each of these methods requires an education process that extends from the manufacturer to the customer.

Finding out what the customer wants and designing a product to meet their expectations is one element of the emerging marketing ideal; so is educating the customer. However, quality cannot be overlooked in an effective sales effort.

Marketing Quality

The word "quality" has been overdefined and overworked until misunderstandings abound. Of the eight definitions cited in *Webster's New Collegiate Dictionary,* the one that best defines quality is "peculiar and essential character." The customer views quality as the ability of the manufacturer and supplier to meet his expectations consistently.

A manufacturer who has diverse raw materials may need to place a portion of seemingly useable material to other uses just to satisfy the customer's quality standards. This may call for extra manufacturing steps such as patching, filling, or packaging. The manufacturer who consistently meets customer expectations may command comparatively higher product prices, and receive more orders in slack markets.

Business-to-Business Partnerships

There was a time when huge single-company manufacturing complexes furnished a seemingly endless array of products while providing employment for thousands. Lumber, plywood, and paper or pulp were the major product lines; Presto logs, closet rods, and garden mulch were reliable sidelines. Most of these manufacturing complexes have vanished, replaced by a network of entrepreneurial manufacturing units that form an interwoven network of increasingly specialized products.

In the face of this specialization, all industry players—manufacturers, wholesalers, distributors, and retailers—need to form business-to-business partnerships. Such alliances help each player to gain a competitive advantage in the marketplace by cooperatively differentiating the products and services each offers. Through these partnerships, individual firms can achieve goals not easily reached by going it alone.

To succeed, these partners must communicate and coordinate with openness and trust. Each party must defer to the others in areas of expertise and specialization,

Exhibit 13.2: Typical Partnering Activities

- Product, service, and marketing programs
- Price, discount, claims, and payment policies
- Logistics, inventory, and transportation decisions
- Advertising and dealer promotion
- Sales force activities
- Manufacturing specifications and packaging
- Communication and site visits between the parties
- Exchange of useful trade and industry information

building on each other's strengths. Together, they define what activities they each must pursue to compete and become more profitable.

These activities range from the joint building of new product and service programs to sales activities that often include the direct participation of key employees. Doing more together, rather than less individually, establishes a powerful structure for providing the "just right" product at the "just right" price, accompanied by the "just right" service—quite a hurdle to clear alone, especially while maintaining a degree of certainty throughout the process.

Outstanding partnering relationships share the following characteristics:

- **Joint planning and performance reviews,** as the parties seek to understand individual and company cultural norms and define the basis and structure for working together.
- **Frequent communication within and between the organizations.** This calls for both openness and a respect for organizational norms, including the guarding of proprietary information.
- **More customized products and services.** When the seller is more involved in the production and design of a product, it improves customer service by allowing the seller to provide not just a product, but also an array of services outside the traditional marketing program.

As the partnership forges a chain of credibility and certainty, the supplier, manufacturer, and seller form relationships that often extend from the forest to the ultimate user. Each participant has a chance to develop additional cost- and value-effective relationships that yield higher margins every step of the way.

In the new business-to-business partnership, the customer plays an active role in defining needs and educating the customer to expedite the process. The new standard also includes the use of "quality" to mean "the peculiar and essential character" of the product that is consistent and reliable. Forest industry participants who embrace the new paradigm will prosper; those that don't will struggle to keep up. The results will be increasingly visible to the outside observer in the years ahead.

14 The Value of Value-Added Products

The forest products industry and its participants have an uncanny ability to overproduce for a commodity market. An unhealthy dose of low prices, accompanied by insufficient demand, usually stimulates a burning urge to increase the value, appeal, or utility of a commodity product, whether it be lumber, panels, or paper products. Add a backlog of high-priced stumpage, and the desire to do something different becomes more than a burning urge—it is a quest for survival.

Operators who become weary of oft-repeated survival quests usually move on to produce value-added products or, at the least, do something to add customer appeal or increase the utility of the product.

Sex and Purple Studs

They aren't exactly purple, but they are bright, with a quality end-wax to protect against high moisture and weather while languishing at the job site. They are Temple studs, known in Texas and other selected markets as quality studs; and as studs go, they are fun to work with. They stand out on the job site, as sexy as a model in the annual swimsuit edition of *Sports Illustrated.* When markets are good, you see more of their plain competitors; when markets are poor, the proportion of Temple studs on the job site increases. The customer is usually willing to pay a few bucks more a load for the Temples.

The forest industry tends to forget the importance of adding customer appeal; they operate as if their products are essential purchases. But people aren't compelled to purchase a forest product. Milton Freidman, the famed University of Chicago economist, put it succinctly while lecturing a group of graduate students: "Food, water, and sex . . . is all mankind needs to survive as a race . . . everything else is a want!"

Forest industry producers occasionally kid themselves into believing that they are performing a vital function. However, their customers have a choice: they can postpone the purchase, reduce the quantity, or go without. The forest products producer also has a choice: continue to produce the same product, in the same quantity, or move on to valued-added and customer-specific products. The producers who move on do so because they're weary of roller-coaster price and demand cycles—cycles spiked by

Exhibit 14.1: Successful Tactics for Adding Value

- Use teamwork.
- Focus on goals.
- Move into value-added products in the good times.
- Know your costs.
- Outsource, if economically viable.
- Invest in adequate research and development.
- Listen to the customer.
- Know the customer's needs.

industry overcapacity combined with seasonally induced demand and periodic economic cycles. Fueled by plenty of creative brainpower, producers who make the move to value-added are writing their own success stories.

Deer Horns and Neon Beer Signs

The more successful and profitable forest industry companies usually pattern themselves after the company model of a Kimberly Clark or a 3M. These companies are rarely heralded in the business columns—but the abundance of Kimberly Clark and 3M products in the home and office speaks volumes. In this business model, entrepreneurial work teams replace the corporate planners, and they have the freedom to design, experiment, and earn rewards for their successes.

A story of new product development and innovation in the old U.S. Plywood organization illustrates the value-added paradigm. Founder Lawrence Ottinger set the pace for innovation and product development. He started off with just a rented storefront, a discovered cache of softwood plywood (purchased for the WW I war effort just then ended), borrowed funds, and confidence that customers would recognize the utility and versatility of plywood in its many forms. His legacy of optimism and innovation became an integral part of the company culture.

The raw-material base at U.S. Plywood's Cascades plant in Lebanon, Oregon, had changed by the late 1960s. Douglas fir three-and-better peelers, the prime raw material, were getting scarce and more expensive. Increasingly, veneer was produced from cull peelers with a rind of clear wood and a heartwood of fomes pini rot, or lacey white spec. The result was a high proportion of A-grade veneer and lots of utility white spec veneer. The A grade was used in the usual upgrade panels; however, the white spec could only be used as interior panel inner plies, with a small part used for the outer ply of sheathing. The excess white spec was all being chipped.

An informally organized manufacturing and technical team, with input from sales and distribution, came together to find a solution to this waste. It turned out to be a panel called Country Place. This three-ply panel had a white spec face, with D-grade crossband and back. The panel was laid up and sized for further processing, and the face wire-brushed to remove the springwood Subsequent manufacturing steps included shiplapping of the long edges, plus the V grooving and staining of the face panel.

The Country Place panel was a winner with the customer—and for the manufacturer. U.S. Plywood developed a profitable use for low-grade veneers; the customer decorated the walls of countless bars, taverns, and family rooms with a rustic, attractive panel. Years later, an original team member reminisced over a tall cool one; he opined that more deer horns and beer clocks were hung on the Country Place panel than on any other panel before or since. He is probably right; and U.S. Plywood's accountants would back him up.

Here are the figures: the Country Place 5/16 inch (8.0-mm) panel sold for about $190 MSM, or about $228 M3/8ths. The price was stable, and rose in a growing market. Compare this with the price of a common three-ply sheathing product, 3/8 CD exterior: during the same 1974–75 time period, it hit a high of $119 M3/8ths and a low of $77 M3/8ths.

Sexy studs and white spec panels are only two examples of what can be done to create value. Adding value to commodity products does take hard work—it takes

Even the lowly peeler core can become a value-added product.

teamwork between the forester, the manufacturer, and technical and sales personnel. It usually needs the unstructured creative freedom of the 3M company model, and it calls for clear goals and focused participation to achieve results. The U.S. Plywood wizardry lay in its entrepreneurial culture, while Temple focused on sales, but the common denominator was a product with customer appeal. Using the same raw material available to competitors, these companies both solved problems and profited.

Avoiding the Misfires

The expensive failures of others provide an insight into developing successful value-adding strategy and tactics. The "what not to do" lessons teach us that value-added now may not mean value-added forever.

During the mid-eighties, paper demand was on a sustained upward curve. Prices were good, demand for most types and grades was excellent, and the industry was ripe for expansion. One large company knew it could bring on a new paper machine more rapidly then its competitors, so it walked its talk, and more—bringing on not one, but several new paper machines. However, competitors were reading the same tea leaves, and had access to the same technology and engineering expertise. They too walked their talk, and a paper glut soon followed. The lesson learned: Don't assume that expertise and market research is one player's exclusive domain, unless patents or other trade protections are involved.

Everybody knew Harry Merlo, then Chairman of Louisiana Pacific Company, had a good idea in the early 1980s. With a pioneering company in Oriented Strand Board (OSB), his balance sheet was better than most. Each new OSB plant added to the total. Then the competitors came along with new plants, some at over a $100 million a copy. Structural panel capacity soared. Production climbed from nearly 36 billion feet in 1994 to 42.3 billion feet in 1997 as 20 new OSB plants were built. And yet some were surprised when OSB became less profitable for most participants.

Louisiana Pacific was not intimidated; they soon launched a program to add value. They developed OSB into products such as soffits, textured paper-overlaid lapsliding and other job-specific items. Profits quickly climbed again, but in the surge of innovation they neglected to carefully define their risk. Claims and lawsuits eventually outpaced the profits. The lesson learned: You must invest in sufficient new product research and development, and technical assistance, *before* introducing products to the market.

The producer can significantly lessen the odds of failure by conducting market research, both formal and informal; performing a technical review of the risk factors; and listening closely to the customers and their decisions.

There are other lessons to be learned. One lumber remanufacturing plant, utilizing lumber previously used as commons, was supplying several Japanese plants with white fir blocks for fingerjointing. At first successful, they soon faced growing complaints, chiefly about dinks—defined as small, sharp indentations on the blocks—and poor grading.

When intensive grading at the plant did not help the problem, the plant manager traveled to Japan at his customer's request. In a loud bilingual discussion at the reject pile, each party tried to outreason the other. Finally, the supplier stopped and asked, "Who is my real customer? Where are the complaints coming from?"

The Japanese managers led him to a grandmotherly inspector on the infeed of the fingerjointing line. A short conversation revealed that the problem wasn't what either the purchaser or the supplier thought it was, because each had a different perception of what the quality of product should be. It wasn't until the manager asked questions about the process and grading specifications and observed the Japanese inspector's activities first hand that he began to understand the problem. The lesson learned:

- Don't assume that people know what they are talking about when they may know less than you do.
- Find the decision-maker and obtain the facts, or at the very least, agree with the purchaser on someone who can fill that role.
- Quantify the facts. Use numbers or visual examples whenever possible.
- Don't rely on words alone to convey meaning, unless the parties agree beforehand what the precise definition of a word is.
- Lastly, work together cooperatively and as unemotionally as possible to ferret out the solution that is hindering the meeting of the customer's needs.

Other value-added efforts have misfired because a company unknowingly stepped into a business for which it was poorly equipped—or made the move at the wrong time. The management culture was out of sync with the mode of operation. The stories are legion; the lessons are real. If you don't know the particular value-added business, don't attempt to run it until you do. Move into value-added products during the good times, when there is ample opportunity. During the slow times, when you have greater competition, it's generally too late to penetrate the market successfully.

Getting the Creative Juices Flowing

Getting the creative juices flowing begins with focusing on a problem, individually or collectively; discussing the potential solutions; and being responsive to the possibilities of new ideas. This process leads to value-added products.

Earthshell Container Corporation, a maker of biodegradable fast-food containers, is an excellent example. Earthshell made its debut on the NASDAQ in March 1998. The initial public offering (IPO) was oversubscribed because of the company's startlingly simple idea: an edible container. This ingenious response to environmental concerns about fast-food container waste certainly sparks the imagination; time will tell whether the idea and product will be a winner. But it owes its genesis to an unfettered creative process.

Dave Pease, in his article "The Last Word: Measuring the Value of Value," wrote:

> Some say adding value to a product can be as simple as marking up the price. It's not as simple as that in the wood industry, where the "value-added" concept has become an article of faith encompassing activities such as product development, downstream manufacturing integration and a variety of marketing and service alliances between producers and customers, to name a few. The concept has grown from an industry buzzword into a movement of substantial proportions. (Pease, 56)

Around 1997 or 1998, Canadian stud producers focused on the United States lumber import quota. This quota, carefully crafted through years of negotiation, protects the American lumber producer from the perceived differences in timber pricing between Canada and the States. That was the idea, but the innovative Canadian stud producers began drilling holes in each stud to accommodate electrical wiring. The resulting "remanufactured" lumber was ruled exempt from the lumber quotas. The initial ruling did not stand, but now the U.S. industry should know that it must keep an eye on the wily stud producers.

Sometimes producers don't recognize the problem and are content to simply wait out the commodity market. A phone call I received from a procurement manager for an overlaid-panel producer is a case in point. The manager was seeking a supplier for southern pine 6-ply 23/32 AC with fewer than nine patches on the face. He commented, "Dick, I can't seem to find anyone that will supply the panel. I have even offered a $30 upcharge and they still turned me down. They won't give me a price."

Patch-free veneer was common in the region, as was the use of the 6-ply construction on the typical layup line. Sure, the glue costs were a few dollars more, and so was the labor. But orders were scarce at the time and prices were low. Even though the upcharge was fair, the producers were unwilling to change. This unwillingness to change blocks the creative process.

However, the industry abounds with examples of companies who do it right. The Weyerhaeuser–Trus Joist MacMillan (TJM) connection is one example. In the mid-nineties, TJM engineers identified OSB as a possible replacement for plywood in their I-beam webbing—if they could find a supplier willing to accommodate the required product specifications on a consistent basis. Tests indicated that open market production did not meet TJM's needs. Enter Weyerhaeuser.

A combined team from the two companies resolved the problem. The solution involved a different adhesive regimen, and board property and test protocol changes. The team got creative, filled a need, and designed a value-added product for both companies.

Some years ago, I was hired to turn a large southern pine plywood mill into a profitable operation. During the first meeting of the plant's leadership, I asked a rhetorical question: "What are we in business for?" The prompt answer: " We are here to make plywood." The equally prompt correction: "No, we are here to make money; we make plywood to make money." This set the new tone for the organization. The business soon made money, and it also made value-added products such as siding and sanded specialties.

The Value-Added Quest for Profits

Most industry members agree that it is getting tougher to produce profits consistently. Hope may lie in new strategic business relationships and innovative goods and services.

More lumber operations are eyeing green lumber not just as a basic commodity product, but as a secondary manufacturing opportunity. The array of opportunities is limited only by cost and the imagination of both the manufacturer and the consumer. Furniture, housing components, pallets, and containers offer just a few of these opportunities. Others include kits of treated lumber for decks, cabinets, windows, and door components. Lumber once considered low-value or no-value can yield value-added products. Board ends can be fingerjointed into specified lengths and then edge-glued into tabletops. All it takes is a little ingenuity.

In this difficult period of industry transition, creativity is the key word for forest products companies. The customer-specific, value-added products they create will benefit both the supplier and the buyer.

15 Forest Certification: Gaining and Accepting Market Access

Some decades ago, building codes and product standards were formulated to ensure that the wood product manufacturer supplied the "just right" product attributes. Consumers needed the assurance of standards for structural and appearance values. They also wanted to know whether the raw material had been grown and harvested in an environmentally sensitive way.

Increased regulations and growing government involvement signal changes. Heretofore, the forest industry has managed its forested lands independently, with a minimum of regulation. This was the enduring industry model—one that the industry would still follow, were it not for the shifting position of the government and the consumer. Their changing roles mandate specific, more visible forest practices that are acceptable to the public as the industry moves forward into the 21st century.

New Realities for the Land Manager

A growing number of critics say the traditional land management paradigm—the forest managed primarily for timber production—is not the best management model. National and governmental bodies are using laws and regulations, and even treaties, to reconcile competing interests. As a higher level of environmental awareness evolves, consumers become something more than just customers. This awareness became evident on the first Earth Day in 1970, and has climbed to unprecedented levels in the 1990s. The increased urbanization of society, media awareness, and the environmental awareness curricula of the schools fuel a growing interest in how the forest industry conducts its business.

A Western Wood Products Association study tried to understand the concerns associated with wood products by those who purchase or specify wood product for others. The study found that nearly 35% of architects and specifiers thought their customers believed that logging and wood product manufacture may be harming the environment (see Exhibit 15.1). The concern with wood consumption extends to its use in paper; some consumers are using as few paper products as possible, thinking that such conservation benefits the environment.

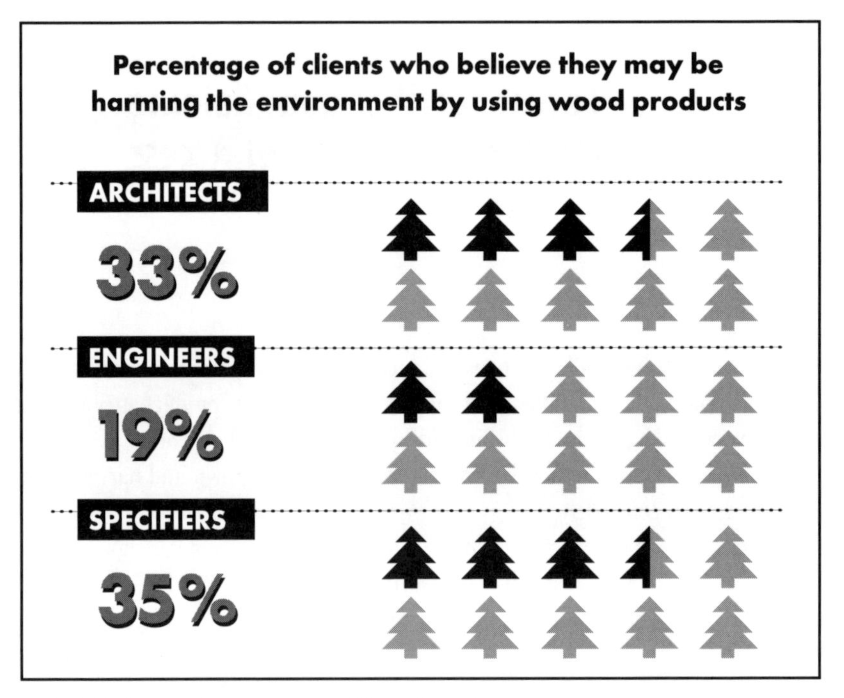

Exhibit 15.1: Consumer Concerns about Solid Wood Products

The forest and wood use debate has shifted from the traditional focus on sustained-yield timber production to an integrated approach. Watershed restoration, flora and fauna protection, and a variety of other interests compete with log production in forest management. Social benefit issues and the rights of non-landowners are gaining stature with government and the courts. Government intervention is here to stay, and so are consumers' needs. Patrick Moore, a cofounder (and now ex-member) of Greenpeace, identified the world's environmental dilemma in a *Canberra Times* (Australia) editorial:

> Many environmentalists seem to forget that there are 5.9 billion humans on this earth who wake up every morning with real needs for food, energy and materials to maintain our civilization. Over the past 10,000 years we have helped satisfy those needs by gradually clearing away about 30% of the world's forests and replacing them with farms and pastures. This trend must now be partly reversed if we want to protect biodiversity and prevent climate change. It cannot be reversed by the idealist notion that if we stop using wood the forests will be saved. (P. Moore 1997)

Those trying to reverse the trend in deforestation have authored additional laws, increased regulation, and new treaties. The Oregon Forest Practices Act that mandates replanting within a year after harvest is an effective and enduring example on the state level. Since its introduction in 1971, it has been a working model for reconciling the needs of the various parties. The British Columbia Forest Practices Code is another regional solution. And there is this U.S. federal government example.

> President Clinton has established the informal interagency Ecosystem Management Coordinating Committee to implement ecosystem management on all, not just federal, lands . . . and the President's Council on Sustainable Development to regulate everything to insure that the carrying capacity of native ecosystems are not exceeded. Both groups advocate strong centralized command and control regulatory systems for management of all U.S. lands and are now firmly entrenched in the bureaucracy. (Adair 1996)

This is only a glimpse of what is happening in a host of countries; increased government intervention is being used as a tactic to gain advantage and effect change. The Climate Change International Summit held in 1997 in Kyoto, Japan, produced a carbon emission reduction agreement, binding upon the signatories, that will mandate reductions in CO_2 emissions as an international effort to prevent global warming.

Today the forest manager must manage the forest not only to supply wood products and paper, but to meet other regional and global needs. Defining new roles for both government and tree grower calls for a new set of precise rules that go beyond the consumer product standards, building codes, and other voluntary and mandatory forest industry sideboards of the past. The consumer is playing a key role—indirectly, through the government, and directly, as a buyer of products.

The Consumer's Emerging New Role

In spite of their greater environmental involvement, governments are often lumped together with the forest industries by consumers as the source of the forest problem. Consumers who no longer see government as the solution are shifting their attention to selective buying and even boycotts of certain products or suppliers, modifying their consumption patterns to achieve social change.

> Traditional institutions and structures that previously acted as sources of security and comfort for society have notably declined in western society. This is perhaps most notable where the role of family and community has diminished . . . with the falling away of family structures, community structures, and other institutions. The individual, by default, has become more reliant on the market as the ultimate social institution to which he or she looks, not only for physical needs but also for social, cultural, and even spiritual needs—a whole spectrum of lifestyle needs of which the marketing profession is well aware. This dependence on the market paradoxically increase(s) the anxiety level. It is therefore not surprising that increasingly in the welfare societies of the First World, price

> is no longer the primary determinant of choice. A complex web of assurance is required from the market, ranging from ethical, animal welfare, health, hygiene and environment. (Viana 1996)

Consumers are using the market to act out their social concerns. These same consumers are convinced that the world's forests are not being properly managed, that trees are being cut and not replaced, and that species are being lost at an alarming rate. They are responding to information that reinforces and promotes what they already believe.

Their perception is valid in one sense: many timber harvests are not supervised by a forester. The Society of American Foresters (SAF), a respected organization well known to the public and the professional forester, estimates this to be true at 60% or more of harvests. These cutting sites, both domestic and foreign, have no forest inventory statistics, no management plan, no designation of set-asides such as steep slopes, streams, and biologically important areas. The SAF is recognizing forest certification as an effective tool for communicating with the consumer and other interested parties. Consumers consider wood products supplied from the certified forest to be sustainable and environmentally friendly.

Verification and/or Certification

Too often, forest owners lack credibility or are otherwise unable to gain public support—and thus many owners, both private and public, are adopting forest certification as a management, marketing, and communication tool. Forest certification, a nongovernmental management model, provides formal verification of agreed-upon forest practices that are incorporated into an overall forest certification plan. The resulting document provides public assurance that a forest is being managed for a variety of values. An effective tool in gaining or regaining consumer support, the ongoing certification process provides a visible assurance to the consumer and others that the land is being managed in an environmentally friendly fashion.

Consumer assurance is fostered through a variety of sources—usually print or electronic media, plus point-of-sale information such as eco-labeling. The reputation and integrity of both the certifying agency and the landowner earn consumer support and acceptance. It's a win-win situation: forest certification assures the consumer that the wood is derived from environmentally compatible sources, and the producer enjoys enhanced market access and a degree of forest practice certainty.

The certification process requires a number of elements to meet the objectives of the consumer, the certifier, and the landowner.

- The certifier must have public credibility and a supplier/consumer focus.
- Agreed-upon certification requirements must be compatible with the law, current regulations, and relevant principles.

- Certification requirements must be recognized as elements of a practical overall system with international standing.
- The certification process must be affordable and accessible to the small and medium-sized landowning enterprise.
- The resulting document must be compatible with the certification requirements of others.
- The certification protocol should be adaptable to different jurisdictions and ecological systems.
- The process should feature a continual-improvement strategy.

Voluntary certification focuses on the highest common aspirations for the forest, while mandatory regulations seek the lowest acceptable criteria. Forest certification will only be successful if there is a consultation process between the parties.

There are three basic approaches to verification and/or certification of forest practices.

- **First Party.** A landowner conducts an internal assessment and practices forestry according to publicly accepted forest practice norms.
- **Second Party.** An assessment is made by a customer, outside trade group, or consultant.
- **Third Party.** An on-site assessment is conducted by a neutral nonaffiliated organization, based on specific norms.

These approaches are described as follows.

First-Party Verification

First-party verification is usually referred to as an internal assessment. The company establishes the specific criteria against which its performance is measured, then periodically assesses and reports, sometimes in the form of a supplement to its annual report. The resulting process is only as effective as the credibility of the company or landowner, and the degree of success in capitalizing on that credibility.

> No matter how well we have done in forestry, we have not given the public the assurance it wants. . . . Our mission is to change behavior and report our results. This will greatly reduce the negative images that have diminished our past good work. The success in accomplishing this objective with reporting internal assessments, rather than certifying results, has been questioned. (Mater 1997)

Because of this questioning, most landowners opt for second- or third-party certification to gain the necessary credibility; however, even the second-party verification is being questioned.

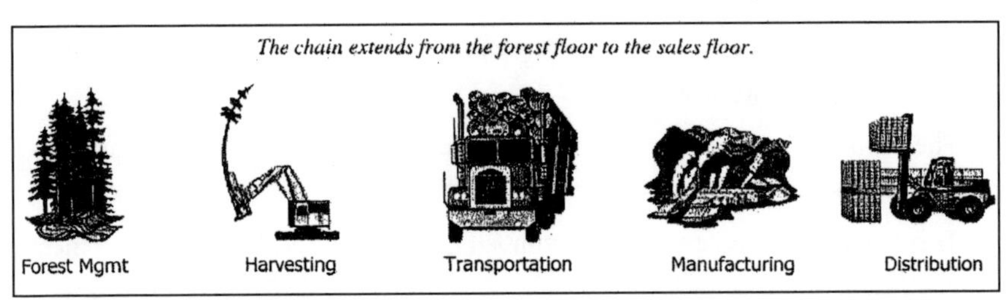

Independent third-party forest certification is becoming a fact of life in the woods and the mills.

Second-Party Verification

The Sustainable Forestry Initiative (SFI) is one second-party verification system. This model was developed and is used by the members of the American Forest & Paper Association (AF&PA) and their contract suppliers. This certification approach is the one most widely used in the U.S. as the 20th century ends. AF&PA members own or lease approximately 90% of the industrial forestland in the country.

SFI has established a set of ten environmental principles, as well as a series of implementation guidelines, objectives, and performance measures. Member companies must report to the AF&PA annually, detailing their compliance. A panel of experts (forestry school deans, public officials, conservation groups, etc.) audits compliance, using data from individual company reports. No onsite third-party inspections are required, although that feature has become an option.

A key feature of the SFI certification program is its national program to train local foresters and loggers in SFI-approved logging practices. The parent organization, AF&PA, carries the SFI message to trade associations, government agencies, logging contractors, state forestry associations, and the public at large.

Third-Party Verification and Certification

Third-party certification is rapidly gaining worldwide acceptance. It is the most credible, from the consumer's perspective. The landowner meets with the independent third-party representatives and defines the principles and practices to be incorporated into a forest plan. The forest owner voluntarily subscribes to a third-party inspection of the forest and an audit of their management practices. This inspection process adds credibility to the forest owner's own reports.

The first certified forest operation was an Indonesian teak plantation, certified by the SmartWood program in 1990. Policies for certification and national standards have since been debated in countries as diverse as Indonesia, Bolivia, Switzerland, Canada, Australia, the United States, and Sweden.

Sweden pioneered the first nationwide certification plan. Early in 1998, Swedish forest companies and environmental organizations agreed on a program to certify that Swedish forest products are manufactured from sustainable managed forests.

FSC, the leading accrediting organization.

Third-party certification is expanding rapidly in the United States and other parts of the world, on both private and publicly owned land. At the end of 1998 there were 11 certified U.S. producers; they will be joined by many others as the trend to third-party certification continues. Like its better-known product-based counterparts, third-party certification is becoming widely accepted by the consumer and the mainstream environmentalist.

The Role of the Accreditor

The Forest Stewardship Council (FSC) has positioned itself as the leading accreditor of third-party certifiers. It is an independent, nonprofit, nongovernmental organization formed to ensure public trust in third-party certification. It also functions as an international evaluator and monitor. Without this umbrella organization or another like it to ensure consistency among the certifiers, the results could vary considerably.

This international organization, now operating out of Oaxaca, Mexico, was founded in 1993 in Toronto, Canada, by representatives from environmental organizations, the timber trade, forestry professionals, indigenous peoples representatives, community forestry groups, and forest products certification organizations from 25 countries. FSC was founded solely to accredit third-party, single-attribute, forest management, and related chain-of-custody monitoring programs. Accreditation is performance-based rather than system-based; that is, it focuses on results rather than on specific performance criteria.

The ten principles of forest stewardship describe the desired end result rather than a prescriptive regimen (see Exhibit 15.2). These principles cover more than forestry;

Exhibit 15.2: The Forest Stewardship Council (FSC) Ten Principles: An Outline of Coverage

1. Compliance with laws and FSC principles
2. Tenure and use rights and responsibilities
3. Indigenous rights and responsibilities
4. Community relations and workers' rights
5. Benefits from the forest
6. Environmental impact
7. Management plan
8. Monitoring and assessment
9. Maintenance of natural forest
10. Plantations and their role

Source: *Certification of Forest Products, Issues and Perspectives,* p. 51.

they also zero in on the social impact of the forest, including concerns of indigenous people and worker rights. Community relations also merit attention. The resulting principles are sufficiently broad to encompass a variety of forest situations, yet narrow enough to provide a framework for preventing degradation of the land. The goal is a framework for defining environmentally responsible management practices.

Exhibit 15.3 summarizes the guidelines to be followed by accredited certifiers in dealing with forestland clients. These guidelines build upon the ten FSC principles and include the basic elements of the relationships among the forest, the client, and other interested parties. The guidelines seek to produce fair, transparent, and effective forest management policies.

The FSC board of directors represents the three areas of economic, social, and ecological interests, with a mechanism to ensure representation from both the Northern and Southern Hemispheres. Formation of national and regional initiatives are encouraged, such as the U.S./FSC initiative. This initiative takes responsibility for public information describing forest stewardship activities, certification, and accreditation within the United States. The U.S. initiative also coordinates the development of regional forest stewardship standards throughout the country, and assists the FSC's secretariat through communication, research, education, and promotion of the accreditor concept.

The accreditor concept is winning support. Private foundations—such as the MacArthur, the Rockefeller, and the Ford—and other organizations provide considerable funding. The Society of American Foresters and similar organizations seek to play a constructive role in defining acceptable forest practices.

Exhibit 15.3: Guidelines for FSC-Accredited Certifiers

1. Compliance with FSC ten principles
2. Independence
3. Sound evaluation procedures
4. Transparency
5. Reciprocity among FSC-accredited certifiers
6. Public information
7. Verifiable chain of custody
8. Compliance with applicable laws and regulations
9. Equity of access
10. Maintenance of adequate documentation
11. Appeal procedures
12. Integrity of claims

Source: *Certification of Forest Products, Issues and Perspectives*, p. 51.

The status of the accreditor is somewhat similar to that of the American Lumber Standards (ALS) committee; it is both the accreditor and the "checker of the checker." Currently, FSC is dealing with the concept of the percent content of certified forest products and how they might be used in conjunction with the FSC logo. Allowing some companies to claim a prescribed conformance level for their products could make the production of certified products more feasible; it could also provide an incentive for steadily increasing certified content.

Market Access

Market access will become increasingly important for the forest product producer, particularly when the vendor and customer have a choice between certified and non-certified sources. The Netherlands is committed to purchasing or importing only certified wood products by the year 2000 or 2001. Other nations, such as Great Britain, Belgium, Austria, Germany, Denmark, Switzerland, Ireland, and France, have picked up the trend. Australia and the United States are close behind.

Forest certification is developing more quickly in Europe than North America, due in part to the influence of buying groups. These groups, which consist of retailers, wholesalers, brokers, distributors, traders, municipalities, and other sellers, strive to supply the customer with certified wood products and also provide assurances by documenting and evaluating their supply sources. Individual participants join the buying groups for a variety of reasons, chiefly

The so-called "big boxes" are chasing consumer demand for certified wood products.

- to act on the business ethic of doing the right thing
- to enhance public relations, or even avoid bad publicity
- to access certified material
- to gain a competitive advantage in the marketplace.

Buying groups, in conjunction with environmental groups, are gaining influence. German publishing companies, which rely heavily on Scandinavian paper products, have fostered an active interest in forest certification in Sweden, Norway, and Finland. It is no coincidence that Sweden was the first country to commit to forest certification.

The WWF 1995 Plus Group, located in the United Kingdom, is a large, well-organized buying group. This group had 75 members by 1997, with aggregate annual sales reportedly in excess of US$2 billion. It is publicly committed to the rapid introduction and sale of environmentally certified products. Among the group's better-known members are such prominent names as J Sainsbury Plc, B&O Plc, Boots The Chemist, WH Smith Ltd, and Tesco Plc. The largest U.K. grocery chain and home centers are listed on its roster of companies.

At least one buying group has been established in the United States to facilitate and increase the purchase, sale, and use of certified forest products. This group subscribes to the third-party forest certification principles of the FSC.

Environmental buying groups are not without controversy; some questions remain unanswered. How do you certify nontimber components such as adhesives, paints, and other materials? What price or product advantage will one competitor have over another?

As the environmental buying group movement assumes a larger role in the marketplace, these questions and many others will need to be answered.

Improved Pricing: The Green Premium

Most forest industry sellers and buyers are still not certain just how much of a premium the consumer will pay for an environmentally certified product. Some say the green bottom line looks good. Others echo AF&PA director of forest policy research Scott Berg, who said:

> This physical disconnect between the grower of the tree and the seller of the forest product makes it impossible for the majority of U.S. landowners to receive a "green premium" or market share incentive, even if one existed. (Berg 1997)

The answer lies somewhere between the two camps. Few consumers consider environmental certification alone when choosing a purchase; quality and availability also play important roles. One definitive study conducted in the spring of 1995 sought to determine the willingness of American consumers to pay a price premium for environmentally certified wood products over a range of consumer products. The resulting paper summarized the study as follows:

> Results indicated that consumer willingness to incur a price premium for certified wood products varied depending on the item considered. . . . However, an average of 37 percent of consumers, across the range of products considered, were not willing to incur a premium for any type of environmentally certified wood. (Ozanne and Vlosky 1997)

The economic and market implications are not yet well understood, nor, for that matter, are the current nature and status of certified wood markets. However, the growing demand for certified wood products has outstripped the supply. These products have found a place in the market.

Green Certification Costs

The cost of certification has received only limited economic analysis to date. There are at least three cost categories:

- The direct cost of certification, such as the application to the certifier, the initial inspection, the annual auditing, and fixed fees such as royalties.
- Indirect costs associated with the first category, such as the incremental costs of improved forest management. These costs will vary greatly depending on the size of the operation under review, the complexity of the forest ecosystem, and the necessary changes to achieve compliance.

  There are some broad estimates of the annual direct and indirect costs. The condition of the forest and the size of the timbered tract will have a significant effect on the cost of certification and the ongoing costs. The actual costs can range from pennies per hectare to more than $2.00. As third-party certifiers and their subjects gain additional experience, both the initial costs and the ongoing costs can be expected to become clearer—and lower overall.

- Chain-of-custody costs, which occur if the manufacturer processes both certified and noncertified raw material through the mill. The costs can be nil, if all certified raw material is processed, or they can be great, given a complex mix of certified and noncertified raw material going through a mill with an equally complex mix of products. The individual landowner and mill operator will predetermine the expected costs before deciding whether or not to certify.

The Eco-Label

Some are promoting the eco-label, or green label, as a product promotion tool to identify environmentally certified products to the consumer. The accreditor's trademark, such as FSC, and the certifier's logo, such as SmartWood, can be used alone or in concert with the manufacturer's label. The goal is to clearly identify the wood or wood-based product as "acceptable." Before long, eco-labeling will become an accepted practice in the same way that the UL label became accepted—and looked for—on electrical appliances in the United States. New accreditation programs, which supplant or supplement existing programs, will spur the growth and acceptance of eco-labeling.

International Organization for Standardization (ISO) and Other Certification

An additional voluntary third-party approach has been initiated through the International Organization for Standardization (ISO). ISO is an international standards body that has developed numerous standards; it is probably best recognized in the United States for the ISO 9000 quality management series. The ISO approach registers a forestry organization to an ISO 14000 series as a vehicle to define and accomplish internally set sustainable forest management objectives. This standard encompasses not only the wood content of the product, but the other materials, such as adhesives and additives.

ISO 14001 is an environmental management system that sets specifications for forest management. The remainder, ISO 14002 and higher, address the environmental performance life-cycle assessment, environmental labeling, and environmental aspects of product standards. They are guidance standards and do not certify the quality or environmental performance of a product. They do certify that an organization is committed to a management process that will enhance prospects for accomplishing both its quality and environmental goals.

The Canadians have responded more comprehensively to the certification issue. They patterned their approach after the ISO 9000 and ISO 14000 standards; however, the Canadian Standards Association (CSA) is a system standard, not a product standard. The CSA standard establishes a process that is structured to lead to improved forest management. These standards, CAN/CSA Z808 and Z809, were adopted in October 1996. While the methods are quite different from those of FSC certification, the results are quite similar, including third-party verification.

The Policy Debate and the Road Ahead

Forest certification as a land management model, to be used in concert with other business tools, such as product standards and building codes, is now an indisputable and significant reality. The policy debate on certification and national certification standards is an active, ongoing process. What is being implemented today will be changed and revised as the players seek better ways to satisfy the demands for accountability and consumer acceptance. What will survive and grow is the whole notion that the public seeks to know and will continue to have a heightened concern for the environment. The consumer, for better or worse, will insist that forests be managed in an environmentally sustainable manner. These certification programs—whether as laws, regulations, treaties, or voluntary efforts—will be used to provide environmental assurances to the consumer and, as such, will become an integral component of forest management.

Section Five:

Moving Ahead to the 21st Century

16 Leadership and Change

Ken Ford was more than a man—he is a legend. The founder of Roseburg Forest Products, his life spanned nearly nine decades of the 20th century. Pappy Ford, as he was affectionately called, started with a single mill that employed 25 people in 1936; 60 years later his company had more than 3,500 employees and an estimated billion dollars in assets. His resourcefulness, toughness, and ability to survive and grow, and his skill in dealing with bankers, unions, and vendors, all made him larger than life.

Like Ford, there are others in the forest industry who were legends in their own time. I think of Mark Reed of Simpson, Larry Ottinger of U.S. Plywood (later Champion International), Bob Pamplin and Robert Floweree of Georgia Pacific, and Arthur Temple of Temple Inland. Later in the century came Harry Merlo of Louisiana Pacific, Red Emmerson of Sierra Pacific, Peter Stott of Crown Pacific, and Harold Thomas and Art Troutner of Trus Joist International. Each seized an opportunity, then spent a lifetime growing, building, and nurturing the vision for what it could be. Thomas, Emmerson, and Stott are still building as the century draws to a close. Other regions, other countries have their own legends—more than a few.

As the 21st century begins, what can we learn from these leaders? What is leadership—and what is not? What business conditions in the 20th century spawned these leaders? How are these conditions changing? What new leadership attributes will be required to be successful in the 21st century?

Leaders, Not Managers: An Industry's Coming of Age

Gene D. Knudsen, a young forester who later became the chairman and CEO of Willamette Industries, observed: "The foresight, tenacity, risk-taking and imagination of company leaders, along with the skills and loyalties of thousands of employees, advisors and supporters, have made the company a success." (C. Baldwin, 1)

Every successful leader had foresight and imagination; they were tough and tenacious, and they knew how to get things done, on their own and through other people. They commanded intense loyalty from their associates, who usually stuck with them through thick and thin, often more than one generation within the same family. Their foresight and imagination offered a vision of what was not yet, but could be.

The tough and tenacious habits of some were described by one industry biographer in the late 1940s:

> Like religious, fraternal, civic and political organizations, the Forest Products Industry . . . has always supported, with tolerance within its own realm, a comparable percentage of sonsofbitches, but regardless of such affliction, the Industry today stands on the threshold of a permanent future. (Cox, 309)

Whether or not a leader fell within the comparable percentage described above is best left to the observer. However, the future was made possible with leaders who were determined and resourceful. They were tough-minded men, driven by a vision; risk-takers who either had an intimate knowledge of the business or had associates who did; men who exploited opportunities rather than solved problems. Each broke from the past, focusing on timber or wood products opportunities and doing something different, something completely new.

What Leadership Is—and What It Is Not

There is little similarity between leading and managing. Leaders create; managers run. Each has a different role; each is important when their roles are balanced within a company. Leadership deals with direction; management establishes structure and systems to get results. One contemporary guru, Stephen R. Covey, suggests that leadership is "making sure that the ladder is leaning against the right wall." Keeping the vision and its attendant mission in sight is the strong suit of the leader. The leader focuses on new ways of thinking about and seeing the world.

Focusing on doing the right things is another hallmark of a leader; the manager, in contrast, emphasizes doing things right. Management focuses on efficiency, logistics, methods, procedures, organization structure, policies, and measurements of efficiency.

You can figure out whether a company is leader-driven or manager-dominated by answering the following questions:

Exhibit 16.1: Core Questions that Define Leadership

- Is there a conceptual framework for the business?
- Do the company communications focus on achievements or previous results?
- Does the business live off the past, with only incremental changes?
- Does the company develop and nurture new processes, products, and businesses?
- Are honest failures and mistakes tolerated in the business?
- Is the company first out with a new product or process, or is it a fast follower—or a laggard?

- Is there a conceptual framework for the business? Is it defined and articulated? If not, chances are a manager is in charge.
- Do the company communications focus on periodic results, or do they highlight achievements in obtaining the vision? A focus on quarterly financial results, or other time periods, is a dead giveaway that the company is management-focused.
- Does it continue to invest in old technologies with incremental changes? If it is still living off the old technologies, it is manager-driven.
- How does it obtain new processes, products, and businesses? If it successfully develops or nurtures them, it is leadership-driven.
- Are failures tolerated in the organization? How is each handled? Eschewing risk and emphasizing control indicates that managers are in charge. Rigid control of product offerings, with little input from the crew, exemplifies the failure-avoidance techniques of a manager.
- Is the company first out with a product or process, a fast follower, or a laggard? The recent history of product, process, and market development within the OSB, LVL, and newer wood/inorganic composite board sectors will single out the leader-run from the manager-driven companies.

Is management then a misdirected profession? Is the management process to be avoided? Absolutely not! A company needs a balance between management and leadership; too much emphasis on one over the other will stop a company in its tracks. But the leader-driven business will be the growth company with the greatest long-term prospects.

The question begs to be asked, "How would these successful 20th-century leaders fare with the coming of the new millennium?" Recognizing similarities and differences between the centuries will help define the answer.

The Year 2001: Beginning a Millennium of New Opportunities

The final decades of the 20th century provide a sneak preview of what's in store for the forest industry. It is exciting! The years ahead will present unprecedented opportunities for the leader who understands the playing field. What are the rules? How have they changed? To what extent is the leader able to influence the playing field and the rules of competition?

The genesis of the new century will be rooted in different ideas and ways of seeing the world. As new ideas and perceptions begin to be recognized and applied, compelling and unprecedented change will begin to unfold.

- **Public Perception of the Industry**

  The community and the consumer will become ever more environmentally aware. The long-held industry norm—that the growing criticism of forest management

Exhibit 16.2: Twenty-First-Century Ideas and Perceptions

- Community and consumers are environmentally aware.
- Consumers are demanding new ways of meeting product needs.
- Outdated industry norms and core beliefs are hindering progress.
- Man is redefining Mother Nature.
- The workplace is a more collegial organizational forum.
- Leadership is being redefined and organizational structure is changing.
- The earth is a small planet.

and the industry practices can be moderated through education—is a futile notion. There are industry practices that the public does not like; the leaders who figure out how to work harmoniously with the public will have a competitive edge.

- **The Forest and the Product: Changing Relationships**
  The long-held paradigm that the tree and the forest drive the industry is shifting to a new paradigm—that consumer demand is the driver—and the new pioneer will find ways to meet that demand, through a variety of raw materials and products.
- **Outdated Industry Norms and Core Beliefs**
  The transformation from machine-based to knowledge-based information is under way. New levels of understanding, and the new theories that flow from that understanding, are spawning new products, new technologies, and new methods.
- **Man Is Redefining Mother Nature**
  The power of the computer to design the tree and design the product is increasing exponentially over time. Those legendary early leaders worked with the premise that Mother Nature gave us what she gave us—we harvested, we milled, we shaped, and we did little else to restructure the resource. Now wood fiber is raw material that is restructured and recreated into products to meet demand.
- **Company and Organization Structure Changes**
  During the quest to get away from the traditional top-down, line/box/layer organizational structure, the forest products industry has tried all sorts of things. The command-and-control mode of the past has generally been eschewed. Industry managers tried to copy the Japanese management methods, Silicon Valley–type work groups, and quality circles—to name a few—with mixed results. The greater successes appear to occur where there is a defined structure yet individuals are treated as associates rather than employees. The company that nurtures leaders at all levels will gain a major competitive advantage in the new century.

The public perception of growing and harvesting timber must change.

- **The Trappings of Formal Leadership**
  The trappings of formal leadership often get in the way of authentic leadership. Not too many years ago, sawmill and plywood managers wore suits to work and were called "Mister." Some managers still do, but although business dress is still important for some meetings, the dress codes for participants at all levels are coming closer together. The suit, tie, reserved parking area, prescribed office hours, and other trappings of formal leadership are disappearing in favor of a more collegial form.
- **Leadership Redefined**
  Leadership is being redefined as "showing the way by going there first." Leadership is more visible, with this enhanced visibility more demanding of individual conduct. Successful leaders lead by infectious examples of integrity, not by charisma. Leaders act the way they want their people to act in the workplace and in the community. Leadership opportunities will be provided to the leader, rather than created by the leader. Listening and inclusion, rather than individual effort and exhortation, will become the tools of the leader's trade.
- **The Earth as a Small Planet**
  With the emergence of the global economy, and ever-fiercer competition for the remaining unexploited resources, the playing field is being reshaped. Trade barriers and tariffs drop; relative transportation costs fall; players rely on common currencies and languages, and on omnipresent information providers and communication devices. Market, competition, and political assessments operate over a shorter time

frame in a dynamic, fast-changing business environment. Risk reduction now hinges on the company's ability to rapidly define, assess, and respond.

An invaluable role model for the new century appeared during the television coverage of the 1996 Atlanta Summer Olympics. World-class athletes demonstrated the concept of a leader mentally creating an action before carrying it out. Viewers could see this as a high-jumper mentally rehearsed the upcoming jump. The athlete's body language, facial expression, and overall demeanor clearly depicted the exact sequence of action, even to swinging over the bar. If the jumper exactly recreated his rehearsal in his actual approach and jump, he cleared the bar. The mental image, the acting out, and the real performance worked in harmony to achieve success.

The forest industry leader will be compelled to use this same technique. The complexity and multiplicity of choices will virtually rule out the "seat of the pants" response. Modeling the changes and opportunities, acting them out mentally and/or with the aid of others and the computer, and then performing similarly in real time will help ensure survival and success.

The New Leaders and the Forest Industry

Twentieth-century forest industry leaders were dynamic, tough, strong-willed and visionary. They could do, they could show, and they could teach. They had intimate knowledge of machinery and people. They knew how to make or sell a product. When they didn't know, they found someone who did. They saw opportunities, then outworked their competition to take advantage of the opportunity. They lived, they built, and they spent decades in the process. They were men of their time.

The new century will require many of the attributes of the past, but with a faster pace, greater knowledge base, and enhanced complexity. Intelligence, information, and language will be the currency of the trade. Leaders will need a much different skill set to find opportunities and respond successfully to them. In fact, an outsider may find it's an advantage not to know what everybody else in the forest industry knows—and not to be burdened with ideas of what can't be done.

Although the United States will continue to have many of the world's largest forest products companies, the major growth opportunities may not be in North America. The dollar may face competition from the new euro as the major trading currency. The common language of trade may be less often English and more often Spanish. The major product innovation and growth opportunities may be in Europe and the Latin American countries.

The American preoccupation with resource and environmental issues; the continued transition from leadership-driven companies to management-driven companies as the industry matures and consolidates; and investors' preference for short-term results rather than long-term prospects; all these will hinder growth and innovation. The industry will grow, and the leadership and growth will be international in scope. The role of any one nation in the leadership and growth has yet to be defined.

17

The Political Science of Forestry: Meeting the Demand for Timber

In bygone days, log suppliers were occupied with growing more and better timber, bidding or negotiating favorable supply contracts, and pushing logging crews. These are still must-do activities, but the task of getting logs to the mill has grown in complexity. Today's foresters, procurement managers, and business leaders are increasingly spending more money, time, and resources in securing a timber supply. Learning from the past will keep us from repeating the past quarter-century of conflict and change.

What are the lessons of the past decades? What are the emerging barriers to a sustainable timber supply? How does a log-based business survive and move forward? What additional knowledge and skills do the forester, the procurement manager, and business leader need? The industry professional can move purposefully forward by knowing the problems faced and the opportunities available; by gaining understanding; by avoiding emotional exchanges with others; and by adapting to the realities.

The Search for an Environmental Crisis

In the spring of 1998, New York's American Museum of Natural History commissioned a Lou Harris poll of 400 scientists and members of the American Institute of Biological Sciences. The poll found that most were convinced that a mass extinction of flora and fauna is under way. They cited human activities as the chief culprits for the expected disappearance of up to one-fifth of all living species within 30 years, chiefly due to global destruction of rain forests and other timber-rich habitats. These scientists believe this poses a major threat to humans in the 21st century.

One participant, a prominent biodiversity expert, claims that species are being lost more rapidly than in any previous historical period—including the extinction created by meteor impacts. Seven out of ten polled believe that the disappearance of species outranks pollution, global warming, and the thinning of the ozone layer as the gravest environmental worry. These scientists, part of a growing academic and lay following, are increasingly convinced that three major reasons account for these alarming developments:

- Unregulated human population growth
- Out-of-control 20th-century technology
- Greedy use of resources by a few

When the public sees polls such as this one, other similar poll results and media stories, natural disasters, and comments from prominent pubic figures, they are spurred to action. The October 1998 issue of the trade magazine *Building Material Dealer* carried the following story:

> . . . members of rock bands Pearl Jam, REM, and Soundgarden have formally requested a meeting with Vice President Al Gore to discuss what they term "the dire state of the nation's forests and wild lands." "This is an issue that hits really close to home for me and the guys," states Pearl Jam's Jeff Ament. "These forests are being irresponsibly mowed down in our own backyard and we feel a need to help stop it and to let people know what's going on." REM's Peter Buck adds, "We can no longer afford to sit idly by as our government gives in to the logging industry. The stakes are too high and the damage already done too great." (*Building Material Dealer* 1998)

These rock stars' involvement provides just one example of celebrities speaking out and gaining public attention. They are not alone; others are as vocal and visible. A natural disaster like 1998's hurricane Mitch, in which thousands lost their lives and extensive media coverage showed the horrors of human suffering, produced another example. In one of the worst summer fire seasons on record for the region, 3 million acres (1.22 million hectares) burned in Central America in 1998. Capping off a long period of progressive deforestation, the summer fires set the stage for the tragedy that followed. The hurricane's winds and rain destroyed thousands of lives and homes in the hilly regions of Nicaragua, Honduras, and adjacent Central American countries. The bare hillsides had little ground cover to moderate the flood that followed.

Mitch was one of the most devastating hurricanes on record. Yet a *Los Angeles Times* article, cited in *Building Material Dealer* (October 1998), attributed much of the storm-caused damage to the massive deforestation that had occurred earlier.

Alarming comments from prominent personalities and natural disaster media coverage intensify the call for action to protect the environment. The National Council of Churches, a representative of 34 denominations, believes churches should consider it part of their work to resolve environmental issues. The council describes global warming as a moral issue "involving Christian concepts of justice." Joan Brown Campbell, the Council's general secretary, stated: "It is often the poor, often people of color, often the developing nations who pay a greater price for environmental waste than others do." (Paquette 1998)

The scientists, the rock stars, the National Council of Churches, and a vast array of environmental groups visualize an array of solutions. Some yearn for a world that is more like the one known in the presettlement era. They seek a simpler, less threatening lifestyle. Recorded memories of centuries past have great spiritual and emotional appeal. Indigenous people are considered role models for living and caring for the earth. Adulation of nature and all living things becomes a religion.

Many share a vision of humans in an eco-utopian future. We would still have computers and other scientific marvels, but these would be closely contained to prevent harm to the environment. Humans would be equal with other living creatures in sharing the planet's resources. We would use our superior intellects to regulate the earth for the benefit of all. And to reach this eco-utopian future, we will need to transfer the wealth and technology from industrialized to nonindustrialized nations, systematically reduce global population growth trends, and reduce consumption of the earth's resources.

The groups championing these beliefs are generally composed of people with high incomes, advanced levels of education, and employment in prestigious occupations in the developed and developing nations. These groups fail to take into account—or don't care—that getting from here to there would encompass a vast social engineering scheme that could seriously disrupt the happiness, wealth, and aspirations of countless millions. Setting the plan in motion would expose the global population to a burgeoning world bureaucracy with increasing jurisdiction over what people say and how they say it. But many with the eco-utopian mindset seem to believe that saving the planet is worth any and all problems this may cause.

Finding Common Ground: Bringing Together the Eco-Utopian and Forest Industry Views

In order to find common ground, environmentalists and forestry professionals must understand the basis of each other's beliefs. Eco-utopians hold several core beliefs, which include the following:

1. Man's new-found technology and emerging culture have created a global environmental crisis that can destroy the natural world.
2. Man's current lifestyle within a fragile artificial world cannot be sustained; it is too dependent upon ever-increasing energy consumption and wasteful technology.
3. The world is a community of nations, each deserving an equal voice and vote in environmental issues and other affairs.
4. The world's environmental needs supercede the rights of any one nation or group of nations (across all political structures, laws, and governmental frameworks).
5. International agreements among the global community members take precedence over other national or international agreements.
6. The United Nations infrastructure provides a base for sponsoring global environmental conferences, focus groups, and initiatives.
7. The United Nations General Assembly is the supreme governing body of the global community.
8. The United Nations, as the world's governing body, is the only effective organization for resolving global environmental problems.

Opinion polls report the extent to which many Americans share these beliefs. For instance, in a May 1995 Roper/Times Mirror survey of Americans, nearly two-thirds of the respondents said that environmental protection is a better choice than economic development. In 1998 an international opinion poll conducted in 30 nations reinforced these American findings. In two-thirds of the nations participating, the respondents agreed, two to one, with the statement: "Present environmental laws don't go far enough."

A procurement manager, tree farmer, or wood products plant manager may ask, "What does a supposed world environmental crisis have to do with me?" The short answer is "A whole lot." A longer answer would cover the following points:

- The world's forests have a huge environmental value for the health of the planet.
- Man's record in caring for these forests is less than pristine, and there are glaring examples of repeated mistakes.
- The public equates clear-cutting with deforestation. To many people, they are one and the same. Until the industry educates the public about this and other silvicultural practices, the public's negative perception of the forest industry will continue.
- The developed nations have evolved from agrarian-based rural communities to industrial-based economies, and then to an information-based era. Most people are disconnected from the land, and unfamiliar with the basics of making such things as sausage or lumber.
- The disconnected tend to ascribe mythical, spiritual, and romanticized values to the land, and demonstrate these values in both tangible and emotional ways.
- The populist view of big companies, particularly huge landowning resource companies, depicts them as despoilers of the land who cannot be trusted.
- Foresters and forest managers represent a benighted profession, using reckless, outlaw methods of forest management and tree harvesting.

Washington, D.C. columnist Larry Swisher recently noted:

> The status of the timber industry has not sunk as low as that of the tobacco industry, but it has followed a similar pattern of industry resistance to facing reality followed by a sharp downward slide of government support. Plainly, its ballroom days are over. (Swisher 1997)

Many industry veterans may still cling to the notion that the "ballroom days" are not over, but they cannot deny the reality of logs that no longer come from federal lands, or changing laws and regulations that hinder, rather than assist, the process of private land management. Most are realizing that the old, "tried-and-true" public relations and lobbying techniques don't work these days, and that a periodic media thrashing is just part of the ongoing business culture.

How then does the industry overcome a tactical disadvantage in the polls, public opinion, and the media? How can a forest-based business survive and grow? Seeking and finding common ground is the first step.

Seeking and Finding Common Ground

The forest industry will not find common ground with its critics solely through an attempt to educate them on the basics of forestry. Even if this could be done, the process is too slow, argumentative, and time-consuming. An effort must be made, but only as part of a greater effort.

The best place to start is by placing oneself in the critic's shoes. The difference in ethical and spiritual beliefs is a study in contrast.

Ethical and Spiritual Contrasts

The forest industry is staffed with business people, foresters, and other professionals. The tools of the trade are forest inventories, growth and yield tables, economic reasoning such as business models, computer logic, numbers, market facts, and technology. The industry's reasoning goes that because it puts five trees into the ground for every one cut, and provides hundreds of jobs, taxes for schools, and an overall economic benefit, the industry should have a compelling public support base. The support base is there, but usually confined to the local community or region. And the story does not sell newspapers or electronic media advertising. A woman living in a northern California redwood tree, risking her life high above the ground in an effort to save the ancient monarchs from the chain saws of a big company, is a more appealing addition to network news. So how does the industry compete for attention and support? By communicating with the industry's critics.

The forester, like any businessperson, tries to form decisions based on the right thing to do; "right" is usually defined on the basis of experience, ethical judgements, economic justification, or other outcomes. The environmental proponent will focus on the moral, ethical, and spiritual dimensions of an issue. The survival of one species may be judged to be more important than the life of a human community. The creature may be gone forever; humans have more options for survival, including government assistance.

Words Mean Something

Industry participants cannot assume that someone using a word understands its meaning or that the meaning is common to all parties involved. The vocabulary of natural resource management and environmental protection is a work in progress. Familiar words such as "forest," "sustainable," "conservation," "ecosystem," "cumulative effects," and "biodiversity" have a multiplicity of meanings. Chances are that current textbooks offer an obsolete or insufficient vocabulary. If the parties hope to understand each other and reach agreement, they must first come to agreement on the technical words to be used, and their meanings. Concurrently, the parties should agree to eliminate emotionally charged words—sometimes called trigger words—from ongoing discussions.

One logger thought that he harvested timber until his teenage daughter asked him if he was a "tree killer." A student thought she was an environmental advocate until she heard herself referred to as an "eco-freak." We all use emotionally charged words to express an attitude, often negative. Attitudes and expressions can spill over into negotiations at the least-expected times, unless we make a conscious effort not to let them.

Civility Is Important to Effective Communication

We may erect the greatest barriers to moving forward constructively when we brand the other party as the bad person. As soon as one party is so tagged, or senses the negative image being portrayed—whether in a meeting, in fund-raising literature, in the media, or in the community—barriers go up. Sometimes these barriers remain for years or decades. As we search for agreement, thinking and speaking well of each other is the first step to bringing down these barriers.

Finding Common Ground

Common ground can be achieved early on if the parties each realize that there is an inherent basis for the conflict, but also an inherent basis on which to build. Each diplomatically contributes by communicating mutual concerns and discovering the topics that both participants can agree upon, before moving on to areas of disagreement. The sawmiller wants a healthy and productive forest; so does the environmentalist. A forester may be an avid fly fisherman; the critic may have a common interest in stream and riparian conditions. The forester, the sawmiller, and the environmentalist each love and appreciate some common benefits of the forest. The process of identifying these interests will narrow down the areas of disagreement and provide a reference point to revisit before moving forward. When we bring a human dimension into the relationship, it's less likely that the parties will reach an impasse. People tend to go the extra mile with someone they know and respect as a person.

Finding common ground may be the toughest task. Unfortunately, for many people it is much easier to disagree and be disagreeable. However, finding common ground is an essential precondition for defining and developing workable solutions.

Defining and Developing Solutions: The Political Science of Forestry

As owners and representatives of the California forest industry gathered at a hotel on the outskirts of Sacramento, California, in 1990, many might have echoed the title of the Clint Eastwood movie *Any Which Way You Can.* They were there to determine what needed to be done to defeat two anti-forest-industry ballot initiatives. Most representatives leaned toward an "off the shelf" political campaign, with the usual fare of TV and radio spots, newspaper ads, and community appearances. Everyone was supposed to dig into his pockets, lend people to the campaign, and get on with it. The industry participants didn't mind underwriting the cost of a glossy presentation by a paid spokesperson, and were willing to personally lobby with public officeholders.

However, some industry professionals thought that relying on that approach was a sure recipe for disaster. Another decision was made, to place a less objectionable initiative on the ballot as a companion to the other two. Without a majority vote, all three ini-

Seeking and finding common ground in the forest.

tiatives failed. The industry then had a breather to work on long-term solutions to the public concerns.

This "any which way you can" political solution isn't reserved for the state of California, or the forest industry either. The forest manager and landowner encounter similar situations repeatedly as the industry critics press their agenda. But they are learning.

Given the local, regional, national, and global scope of the forest industry's critics, they need to learn fast. A consortium of industry critics has been pressing for additional wilderness and less-managed forests for years—and they are good at it. In some regions the annual harvest from federal lands is down to 10% of the level in prior years.

The following are examples of what the industry faces:

- A National Environmental Education and Training Foundation poll shows that people increasingly support environmental protection even when ignorant of the issues. Poll participants were asked to answer 10 multiple-choice environmental questions; correct answers averaged 2.2 of the 10. Random guessing would normally average 2.5.
- The public will not accept harvest practices that one writer says will "cause the earth to look like a great wounded mother, destroying her beauty in the vicious aftermath of clear-cutting." ("Globalization: Surfing the FiberNet" 1996)

If "any which way you can" politics is one effective political tool, what are some others?

1. **Know and understand the tactics and objectives of local, national, and global environmental organizations.** Be informed, and share with others. But don't just look for a "logs versus the environment" agenda. Many industry critics seek political power and control, at the regional, national, and international levels. Enlist the aid of others outside the industry in dealing with forestry and forest-related topics. When seeking allies, consider political power and control issues as core topics that may overshadow forestry issues in importance.
2. **Don't feel picked on; everyone gets a turn.** Environmental concerns are here to stay, and may be expected by any party building a power-generating facility, a new factory, or even a home. Some have discovered that land improvement can turn into a nightmare of bad publicity and permitting headaches, as the following two examples illustrate.

   When Hyundai proposed siting a computer chip plant in a western Oregon community, the plant and its 2,000 jobs were fought as an enemy of nature because the facility was to be situated near wetlands.

   In the Libertyville Township of northern Illinois, an 80-acre cornfield on Butterfield Road was restored to its natural state for $1.2 million, "the way it was when Pottawattami Indians still roamed the area." The township will sell mitigation rights to builders in the Des Plaines watershed so that the latter are allowed to build homes.

   Some say that we have created a "conflict industry"; that resolution of one environmental issue leads to another. More issues, and more successes, equals more donations and grants. This process provides employment for the nongovernmental organization (NGO) worker. One innovative publisher sells an annual directory that identifies 750 environmental grant-making foundations. For $94, plus $6 for shipping and handling, the grant warrior can complete a number of applications that are tailored to the grant-giver. Direct-mail appeals and sales of T-shirts, books, calendars, and other souvenirs help fund the effort. Occasionally there is an angel giver, an affluent someone who donates directly and has great influence. But nothing works unless there is an issue, and a track record of success.

Faced with the magnitude of opposition, the forest industry would do well to seize the initiative with its critics and work cooperatively to resolve issues.

3. **Beware of government initiatives.** Don't expect governments to benefit the forest or the industry. They seldom do. Be surprised when it occasionally happens. The reality is this—governments respond to political pressure over the short term, but forest management is a long-term business. Money spent on the forest gains few votes in comparison to social expenditures. As a result, government forests are usually shortchanged.

   A June 1997 story published in a Japanese newspaper, *Nikkei Weekly,* provides an example. The staff writer described the dire fiscal and ecological shape of Japan's national forests. He cited poor management practices and chronic money problems. Some of the national forest offices could not even pay salaries or light bills on time. Planting of the wrong species and a variety of other mismanagement practices added to the woe.

   Look for rhetoric, not performance, when a government manages forestlands. Participate and try to influence improved performance. That's the best the industry participant can hope for.

4. **Unify, and press forward as an industry, within the laws and customs of your nation and that of other lands.** Look for allies anywhere they can be found. As the parties come to clearly understand the real issues, each controversy will attract allies. The Sustainable Forestry Initiative (SFI) is one example of what to do right.

   The American Forests & Paper Association (AF&PA) founded the SFI after it conducted a 1992 public perception survey. Industry CEOs saw the problem as a lack of public knowledge; others in the business believed the industry had a behavioral problem. The CEOs did further research and came to the same conclusion. The SFI was tailored, communicated, and implemented to deal with behavioral problems.

   Gaining commitment and understanding from the AF&PA membership and their suppliers was a huge task. Members and suppliers who would not make or honor commitments were dropped. It was a courageous move, in the best interests of the industry. It identified acceptable practices and encouraged members to do more, and most did just that—gaining allies in the process.

5. **Use past mistakes as examples of what not to do.** Learn from your mistakes and those of others. There are lots of mistakes and they are visible. Be open with your appraisal of past mistakes. Talk about what has been learned, what is being done to

mitigate the mistake, and what you are doing to prevent reoccurrence.

6. **Know what you stand for.** The mission statement of the Society of American Foresters, and the position taken on seven controversial issues, showcases the best professional thinking. The mission and position statements were adopted with 91% approval in December 1995. Divisive issues such as clear-cutting, use of chemicals, and fish and riparian considerations were addressed. The industry should periodically review and renew communication of these statements. The review process should not be reserved for the forester alone; it should include the media and others inside and outside the company.
7. **Walk your talk.** A trade group's actions represent the industry; so do the actions of single companies and individuals. Unfortunately, a black eye for one is a black mark for the industry. And it is possible to get both a black eye and a black mark, even with the best of intentions. In late September 1998, the *Seattle Times* front-paged a six-segment series entitled "Trading Away the West: How the Public Is Losing Trees, Land and Money." The first segment examined the recent government land trades with private companies. That one grabbed the reader, even though the trades were in the best interests of both parties. The next segment dealt with a land speculator, not connected with the industry, who had purchased a private tract in a Colorado wilderness and begun construction of a home. Construction activity, the use of a helicopter, and announced plans to mine for coal brought a land trade offer too good to refuse. An even-steven land trade with the U.S. Forest Service netted the land speculator about a $ 3.5 million profit within 18 months. But who got the recognition in the public's mind? The forest industry.

   However, the industry got a different kind of credit when Champion International and International Paper sold land to the Nature Conservancy in the northeastern United States. This is image-building credit, and a win-win situation for both the public and the companies involved. MacMillan Bloedel gained more credit for the industry when it agreed to modify its harvest practices to foster improved relations with its neighbors, customers, and other constituencies. Survival politics require the industry to police itself, walk its talk, and proactively meet the real needs of its public.
8. **Be alert for unexpected opportunities.** The 1988 Yellowstone National Park fire was a tragedy. Even more tragic was the criticism and second-guessing that followed during the subsequent years. The Yellowstone fire is requiring land managers and the public to reassess the role of fire in the forest.

   Other activities—such as species selection, harvest practices, management of riparian zones, and a host of other land management practices—are now being examined and reassessed. Perhaps the 21st century will usher in the Age of Discovery, as we evaluate and learn from nature and controversial events. Sharing the learning experience with an interested public is essential.

9. **Take the lead in defining environmental objectives.** The company should prepare an environmental management plan as a companion document to the forest management plan, and share it with the public. Plum Creek Timber Company, located in Washington and Montana, is getting it right. Over the years this company has repeatedly drawn the unwelcome attention of the environmentalists and their listening audience. But lately, Plum Creek has become proactive in changing, communicating, and moving forward to improve forest practices and their image. They call it Environmental Forestry, which includes environmentally sustainable forest management practices. An annual third-party audit verifies compliance. Results are published in Plum Creek's Environmental Management Report. The company even has a third-party auditor annually assess environmental compliance at its manufacturing sites. Performance audits and other environmental-specific information are available through a website.

   Plum Creek is not alone in this aggressive push for openness and good citizenship. But the company represents what is happening within the industry to provide a foundation of understanding with a company's critics and other communities of interest.

10. **Allow public participation in forest planning, even on private lands.** A forest management plan can be best shaped with the input of others—sometimes even with a company's worst critics. Find interested individuals that are respected, have interest, and can commit to a group process.

    MacMillan Bloedel has asked representatives from labor, the British Columbia provincial government, native Americans, local communities, and environmental groups to review the company's forest management plan and provide recommendations for improvement. These meetings are structured to promote discussion and interaction between the parties. Each meeting is structured to define the task to be accomplished. All agree to avoid philosophical discussion and ill-defined meeting goals. MacMillan Bloedel decision-makers attend each meeting to ensure that the participants understand that it is a working session and results are expected. Outside contributors feed information and challenge the group. The temptation to reject ideas out of hand is deftly handled; each idea is examined to determine its unique possibilities. The process develops a management plan that is doable and acceptable.

11. **Accept third-party oversight.** Obtain a stamp of approval from an independent auditor, such as described earlier at Plum Creek or as described in Chapter 15, Forest Certification. This assures interested parties that the company is conducting sustainable management practices, and has been certified or verified by a third party acceptable to a consumer group and the public at large. MacMillan Bloedel, Champion International, and Collins Pine are some North American companies that have chosen third-party oversight. Third-party certification or verification is better than just good publicity. It's a tangible way of demonstrating that the company is committed to sustainablility.

12. **Do something about the education of the children.** Three generations of Baldwins have earned timber dollars over six decades in communities where timber-based income was a major economic factor. One evening I observed my daughter preparing a large multicolored drawing of a regenerating forest stand. A written report was to accompany the drawing. I reviewed it and asked questions. What followed was a surprise. Cutting trees harms the land—at least that's what she believed, based on the instructions provided by her biology teacher. I immediately rescheduled my coming Saturday. We went to the woods and walked the land, and viewed what was actually going on.

    Unfortunately, only a few of the new generation of children and young adults have the time or opportunity to really understand forest management. But the industry needs to commit time and resources to the children, and their parents. The schools need a teaching aid such as *Project Learning Tree,* a publication of American Forest Foundation.

    Project Learning Tree provides a balanced view of supply, the use of trees, and environmental considerations related to the forest. This publication should be an essential pre-K to grade 8 learning tool. Parents, using this as a home teaching guide, will gain in knowledge as well.

The Political Science of Forestry is a new science for the new century. Forestry, wood products, and wood products marketing are encountering growing challenges to doing business as usual. The industry needs to catch a vision—a vision that says that we can have forests, we can have a forest-based industry, and we can have a multiplicity of products from the forests. And wood, a whole lot of it, can sustainably come from the forest. The vision needs to be shared with others. Cecil Wetsel, Jr., and his wife, Judith, know all about that.

Wetsel-Oviatt Lumber Company, situated east of Sacramento, California, is a second-generation family business. The sawmill, aggressively retooled in recent years, continues to compete for trees, labor, and other resources. The surrounding community, once a rural agricultural-based economy, is now a bedroom community for the high-tech industry. Cecil once considered becoming a preacher; he may have found his calling running a lumber operation.

> We need to let the country know that when they're using wood products . . . it is coming out of a renewable resource. . . . And that is going to be the rest of our lives—to spread that and communicate that and have people talk about that. (Benson 1998)

The whole industry needs to do that—commit itself to communicating a vision of a working forest, working mills, and customers that like what the industry does. That's the Political Science of Forestry.

18 Business and Product Planning: Getting Inside the Competitor's Decision Loop

Capital budgets were tight in 1984, but everything told the company I worked with that going ahead with a new peeling concept—a concept utilizing low-cost pine pulpwood as raw material—would be a winner. The primary questions were: "Will the process work? What happens if it doesn't? Will our manufacturing costs be low enough to make a competitive difference? Will the equipment materially expand the wood supply available to the mill?"

A colleague then made a statement that brought all the questions into sharp focus: "Dick, if you are on the wrong train, every stop is the wrong town!"

As we asked, and then answered, these and other questions, everything told us we were on the right train, heading to the right towns. That's the end result of business and product planning: picking the right train and the right towns that follow. But just how does a business pick the right train? What is the process? What is the end result?

Pick the Right Train

Up to and into 1984, the competition for timber was heating up, along with the rising demand for plywood and lumber. Mills were closing at an unprecedented pace because of a lack of affordable logs—about 20 mills each year in four Western states alone. Georgia Pacific and Louisiana Pacific decided to do something more than incrementally modernizing their existing mills. They opted to make major investments in a new product, Oriented Strand Board (OSB), when others were pulling back from the structural panel sector.

> Georgia Pacific Corp., which cornered 40% of the southern pine plywood business by moving into that market early, is setting the pace in the new composite panel business as well. It already has two plants in production, two under construction, and six more planned in a $400 million program emphasizing high-strength strand board. Tactics at Louisiana Pacific Corp. call for a larger number of smaller plants turning out lower-strength waferboard. Chairman Harry A. Merlo calls his product "the smart man's plywood" and vows to be the largest producer of the new generation of composite boards. (*Business Week* 1984)

By year-end 1997, Georgia Pacific had six plants onstream producing 1.6 billion BM 3/8th; Louisiana Pacific had fifteen. Harry Merlo's vow had become a reality: LP was the undisputed top producer of composite boards, with in-place capacity of 4.8 billion BM3/8th.

Although each company subsequently went through management changes, each is a world-class company that has the demonstrated ability to identify trends, think strategically, and implement the business plans. Each company has a history of picking the right train.

An article in the March 1989 *Forest Products Journal* described a study prepared by Bilek and Ellefson at the University of Minnesota. They identified and listed 26 publicly held forest products companies and researched the stock ownership, and found that:

> Ownership of their own stock was not common among the wood-based firms studied. Only 4 of the 26 chose to do so (Georgia Pacific, Kimberly-Clark, Louisiana Pacific, and Masonite). In each of these cases, however, the ownership placed the company among its leading three shareholders. All such ownerships were for employee stock plans or employee incentive programs. (Bilek 1986)

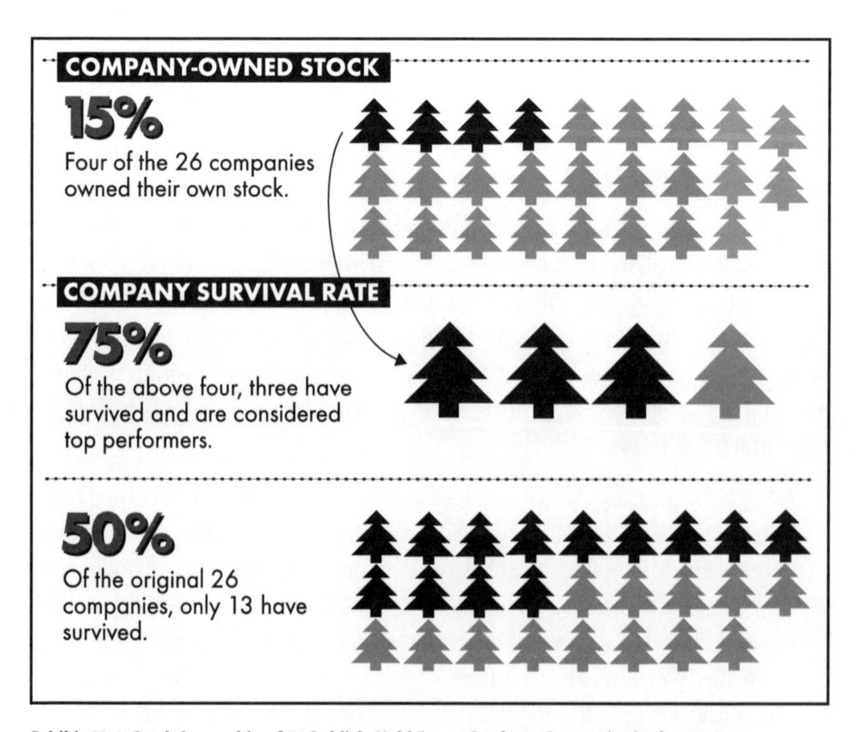

Exhibit 18.1: Stock Ownership of 26 Publicly Held Forest Products Companies in the 1980s

The results over the remainder of the century are equally startling; three of the four (GP, KC, and LP) not only survived, but are considered top performers in the industry. Of the 26 companies in that article, only 13 continue as stand-alone companies into the new century (see Exhibit 18.1). A 50% survival rate over a 15-year period consigned some well-known names—such as Crown Zellerbach, Diamond International, St. Regis, Hammermill Paper, and, lately, Union Camp—to the history books.

Plans Made, Lessons Learned

The forest industry participant can learn a number of business planning and plan execution lessons from events of the eighties and nineties. Here are a few.

Business Planning Is a Continuing Activity

Georgia Pacific had hit it big in the west during the fifties as a young, aggressive company. Beginning in Washington State, GP purchased mills on the cheap, modernized them, and skillfully negotiated a timber supply. In the process, it billed itself as the Growth Company.

Then along came southern pine. Timber was cheap and plentiful in the southeast, and the major east coast markets for western plywood were days and dollars closer.

The longer-term outlook in the west wasn't nearly as favorable. Planning, research, and studies followed. Questions arose, such as, "Can we peel small-diameter logs and still get production? And can we bond the veneers of southern pine logs with hard-to-glue features of wide summer-wood bands and high resin content?" The rest is history.

GP took the lessons it had learned at Springfield and Coquille, Oregon, at other mills, and at its laboratories and transferred them to Fordyce, Arkansas. Georgia Pacific's march across the south began. Fordyce became a "cookie-cutter" for a dozen mills and more. Sales soared, and so did profits.

Given GP's history as an innovative, fast-paced growth company, the announcement of the company's move into OSB was an expected event.

The lesson learned: Once is never enough. The forest industry is a fast-paced, ever-changing business—and the process of business planning and strategic implementation never lets up.

Intelligence and Information: Reducing the Risk in Risk-Taking

Information is the lifeblood of the business; intelligence, accurately interpreted, provides insight into what should be done over the short and the longer term. Acquiring information takes a massive effort, an effort that looms ever larger as the forest industry becomes increasingly global. It requires the building of formal and informal communication networks that span the globe. It also requires understanding of all parties' behavior of the parties—not just the organizational structure—that can affect your business interests. By reading intentions and reasons, and not just actions, sometimes you really can predict the future.

Georgia Pacific, Louisiana Pacific, and other forest-based companies had access to the same information during the 1980s. The big logs of the west were just about gone, while western plywood production costs were continuing to increase. Plywood manufacturing costs in the southeast were also rising. The markets for structural panels were growing and would continue to grow as the Baby Boomers became home buyers. The composite board technology was no longer an infant, and advances in engineering and processing enabled the production of a plywood substitute at a fraction of the cost. Other panels makers were pulling back in favor of increased investments in papermaking capacity.

The resulting intelligence was quite similar. Now was the time to make a major play in composite panels. Louisiana Pacific and Georgia Pacific did; others didn't. Some didn't because they disliked the risk. One company had had a bad experience with a plywood plant some years before and felt that the greatest opportunity would be in southern pine lumber. Even Weldwood of Canada, which had suffered through a decade or more of product development on two first-generation Canadian waferboard plants, chose not to compete further.

The lesson learned? Gathering intelligence in a global economy is growing more complex and time-consuming, but it is a crucial element of business planning. "Reading tea leaves" is an art that requires understanding behavior and intentions before acting; and it requires risk-taking and judgment calls.

Not Having a Plan Is a Plan

An old proverb states: "Whom the gods want to destroy, they send 40 years of success." In the 1980s, the gods sent Sir James Goldsmith, too. Sir James didn't know a whole lot about trees, or papermaking, or wood products. But he could smell value and underappreciated assets. He was also a strategic thinker—with access to capital.

Crown Zellerbach, Diamond International, and St. Regis were more than 40 years old and considered successful—well-known public companies with enviable reputations for quality products and stable operations. What little strategic thinking went on there probably occurred at well-staffed corporate headquarters, or in the mind of the occasional maverick in the field. The businesses generally acted the role of big, stodgy corporations—which they were. They left themselves open to a value finder, such as Sir James.

Goldsmith acquired Crown Zellerbach and Diamond International, broke them up into more valuable segments, and subsequently sold each segment in a timely fashion. St. Regis, fearing the same treatment, fled into the arms of a white knight.

The lesson: If you have a valuable company, keep in mind that there will be outsiders doing strategic thinking and business planning for your company.

The Right People Do the Planning

Some years ago, a division wood products manager was directed to work with the corporate planning department at the company's eastern headquarters. The manager answered questions, made recommendations, and provided staff-type assistance to the young MBAs in developing a five-year strategic plan and the coming year's tactical plan—strategic visioning, it was called. The resulting plan—the latest business-school textbook solution—was folded into the overall company plan. However, the narrow input, the rigidity of the process, and the proposed solutions offered nothing new or unique. It was just a plan, like any other industry plan of the time—mostly an extrapolation of the past.

A business plan is a dynamic creation. The guy in charge articulates the vision, critiques the process, searches for alternatives, and pushes the plan to completion. The plan is not merely a byproduct of the business; it requires the total focus of the company. Inclusion is created at all levels. The resulting team is a blend of experience and vision, a look at where the company fits within the industry. The plan defines the profitable prospects for the business. It is a working document that stakes out the future for the company. The professional planner (plus a supporting player if one is needed) guides the process, provides structure, plays the role of devil's advocate when needed, and sums up.

Lessons learned: The responsibility for the plan belongs to the top guy. Active participation by all levels—involving a blend of experience, business acumen, and forward thinking—should result in commitment to the goals articulated during the planning process.

The Planners Create a Vision

Harry Merlo had a vision; he knew what he needed to do. He wanted to be the top producer of the "smart man's plywood." A later idea was Nature Guard, an insulation product manufactured with material gleaned by "harvesting the urban forest" of recycled newspapers. He wanted to be a friend of the builder and homeowner by providing affordable building materials. And he wanted to make lots of money for his company. "Merlo-watching" became an investor and competitor preoccupation. His Louisiana Pacific theme of the seventies and early eighties, "YES WE CAN," helped spread his vision throughout the company.

Anyone evenly remotely connected to the industry was aware of the Georgia Pacific vision; they were the Growth Company. "The Growth Company" was posted on their products and their business communication; it was always referenced in their print and electronic media. Similarly, even those outside the industry know the Tree Growing Company, Weyerhaeuser. Mention the name; tree-growing immediately comes to mind. Weyerhaeuser builds mills to use trees, makes products to use trees, but most of all, it grows the trees.

Each of these companies is an industry leader; and each has a vision. The vision is the driver in strategic planning, and in the manner in which the business is conducted overall.

The lesson: If a company doesn't have a vision that can be defined in a multisensory fashion, it needs one.

Capital Expenditures Are Proportionate to the Task

Too often, the term "strategic planning" is just an annual code word for capital planning. The question often comes up early on in a planning meeting: "How much do we have to spend this year?" Further planning then seeks to justify the capital spending.

While capital planning is important, it should be a solution for positioning the business, not an end unto itself. Capital planning should be judged on getting the "biggest bang for the buck." Peter Drucker, considered by most to be the founding father of modern management, frequently mentions paper-mill capital spending as an example of what not to do:

> ... the trade-off of capital for labor has not always worked. One example is the paper industry. ... [W]hile there has been a very sharp increase in the productivity of labor in papermaking as a result of large-scale mechanization, it has not been nearly as great as the decrease in the productivity of capital. The trade-off has not worked. (Drucker 1980)

A newly installed paper machine costing a billion dollars or more is a common paper capital expenditure. When asked how the huge cost is justified, owners answer, "It's a long-term business; we are positioning for the future." How about the $50 to $100 million lumber operations of late; or the $100 million-plus composite board plants?

Lessons learned: Capital spending is a tool of strategic planning and not an end unto itself. The more efficiently the money is spent, the greater the return, and the lower the risk for the investors' dollars.

Listening, Inclusion, and Incentives Are Tools of the Planning Trade

Listening and inclusion are really tools of the trade. They provide a diversity of ideas and observations, most of which aren't readily available within a select group of managers and planners. Listening and inclusion are the foundation for commitment to a course of action. Add incentives, and committed individuals become stakeholders.

The top-performing forest products firms, such as Kimberly Clark, Louisiana Pacific, Georgia Pacific, and Trus Joist, offer stock ownership plans to a cross section of employees. These companies are discovering the value of employees as stakeholders.

Lesson learned: Employees that are stakeholders provide an added level of commitment to the company goals.

The eighties and nineties will be remembered as decades of unprecedented change and restructuring. There was a high casualty rate for both plants and multilocation companies. The cause and effect of these casualties provides many lessons for the survivors, and for new entrants in the industry. However, companies change and so do their people. The lessons learned by the surviving companies may not be institutionalized sufficiently to keep them from losing their bearings. Look for more surprises, and casualties, as this industry moves forward into the new decade—driven by proactive companies playing by their own rules and driving their competitors crazy.

Play by Your Own Rules, Drive Your Competitors Crazy

Interpreting events through insight and foresight, anticipating the emerging opportunity, getting ready by building capabilities, and ignoring conventional wisdom: these are prerequisites for building a business. Getting there first is a great way to move forward. The June 1998 Oregonian article on Crown Pacific, the sixth largest private owner of Pacific Northwest timberland, is case in point:

> Crown Pacific Partners, founded in 1988 with $1 million from Stott [Peter W. Stott, president and CEO] and $35 million from banks and other investors, is the stealth bomber of the Northwest Timber Industry. Stott and his partners saw opportunity while others saw trouble. They borrowed heavily to buy timberland at a time when increasing environmental regulations were driving sawmills out of business. They relied on business smarts, not conventional thinking. (Brinkman 1998)

The company, reporting 1997 revenues of $505.6 million, has enjoyed extraordinary growth and profitability during its first decade. Peter Stott had a vision.

> We looked at a industry that was beginning to change and saw a niche for ourselves. . . . I had a fundamental feeling that the value of wood fiber would appreciate over the next decade. It did. (Brinkman 1998)

Interestingly enough, Crown Pacific's first deal was for a central Oregon timber tract owned by Sir James Goldsmith's group. Stott and his partners assembled the tea leaves and read them. They developed a strategy and Crown Pacific moved forward, playing by their own rules.

There will be other leaders and companies with visions. Some may be as successful as Stott, and some will fail. But the successful firms will have characteristics in common.

- **Impromptu Planning Sessions**
  More often, less formally, and with the right players in attendance—this is an effective way to do short-term tactical updates and long-term planning. Sam Walton's regular Saturday morning meetings established a model for Wal-Mart and others who seek visionary growth. Wal-Mart's top people were required to attend Saturday morning meetings where the weekly results were critiqued, the game plan reviewed for the following week, and the longer-term plans developed. It wasn't a formal affair, and it tended to be a bit wacky at times.

Who could call a meeting serious when the company chairman started the meeting with the University of Arkansas cheer? Some might say that "calling the Hogs" is not exactly the proper way to start a meeting, but Sam thought it was. He figured it was the best way to wake everyone up. Sam's famous hula dance on Wall Street was an embarrassing, but profitable, outgrowth of a lost Saturday meeting bet.

Walton was an early riser; he read the numbers as a premeeting ritual. This further established the example for others. Performance critiques, evaluation of competitors, things to do immediately, and future plans—all were grounded in in-house facts, other information, and intelligence from the field. The meetings were fun, but also vital planning and strategy sessions.

Whether on Saturday or some other time, the successful forest industry competitors will have meetings, both serious and fun. Active participation of the group members will step up the company's tempo—and it will respond to new opportunities and circumstances with greater agility.

■ A Sense of Urgency

When I was a boy growing up in the Willamette Valley in the 1940s, it was the custom for all the family members (who manned the family-based floral/nursery business) to gather for lunch at a huge kitchen table. Grandfather sat at the head of the table and the greater number of his six sons would attend, as occasion and the war permitted. The working grandsons, myself included, occasionally got to sit at the table in recognition of their work. This was the small-business, single-family version of the Sam Walton Saturday meeting. But there was a marked difference: the plans made seldom got translated into action. Yes, there was the distraction of World War II, and there was no grand design to become a world-class company. But Grandfather's business, then the largest floral business in the region, later became a site for a modern retirement home.

Effective meetings will focus on growth in sales and profitability, and ways to achieve these. Walton's meetings focused on plans, implementation, and expected results. The tempo was upbeat and the sequence was repeated weekly.

Seizing challenges, jumping curves, and shifting the playing field give a company a competitive advantage within the decision loop. In the end, the planning meeting should position the team and the business to do just that.

■ Strategic Planning and the Competitors

OSB plants aren't for everyone. The early entrants, after the usual debugging of equipment and gaining of customer acceptance, were the most profitable ones. The first followers generally did well. However, the role of each of the participants is a good case study in how to compete, or not compete. The ones that drove their competitors crazy did best.

These companies were the ones that chose to compete at all levels, from chairman to hourly employee. Once they saw the opportunity, they moved quickly. Some of the projects didn't work out, but enough were successful that total investments performed well above average. The companies identified needs; they brought customers into the process, and made some of them evangelists for the product.

Every face of the OSB panel had the characteristic thin wafer flake design: no knots, no splits, or any of the other visual defects typical of a CD grade of plywood. The resulting OSB faces were identified as A grade, a designated appearance grade in plywood.

Products for niche markets soon followed, replacing plywood and other products. The competitive advantage came through *transforming,* rather than refining, the product in relation to its higher-cost competitors. The first few OSB marketers literally owned the trend, as OSB production and sales climbed to 40% and more of the structural panel market.

Are there business growth opportunities out there? You bet there are. They may emerge as products with never-before-used raw materials, products that have a unique consumer appeal, or opportunities yet to be discovered by a Merlo, Stott, or Walton. One thing is certain—the opportunities are there, if imagination and resourcefulness are too, along with a vision that some call a business plan.

19 The Achilles Heel: Demand Versus Expectations

How does the forest industry balance product demand with the consumer's broader expectations? Some say this can't be done without turning back the clock on timber supply decisions, but mounting evidence refutes them. The preceding chapters have provided an overview of changes that have impacted the industry; how the land manager, the manufacturer, and the consumer have responded to these changes; and what we can expect in the new century. We examined raw material choices, conversion options, and market opportunities. This conclusion will highlight those changes, spotlight areas of opportunity, and summarize emerging trends for the decades ahead.

An Overview: Where We Are, How We Got Here

Some landmark events are singular catalysts, sparking a sea change in power, process, and pattern. The result is a paradigm shift. That probably isn't what Judge Maxwell had in mind when he enjoined the U.S. Forest Service from carrying out a plan to clear-cut the Monongahela National Forest in West Virginia, but his December 1973 decision set in motion forces that would forever change the face of the forest industry.

The changes would transcend the boundaries of the United States. A domestic outcry by the Isaac Walton League, Sierra Club, and others initiated the Monongahela lawsuit. Similar outcries in other nations would grow in intensity and emotion. Initially, the Forest Service and forest industry claimed there was no real problem, just a lack of understanding and education. But we now know that a fundamental shift in public sentiment had indeed occurred. This paradigm shift, ignited in its infancy by special-interest groups, set in motion forces that have fundamentally changed the public perception of forestry and the forest industry.

The vision of an industrial-based future, with production and economic decisions as core values, gave way to an information-driven, computer-based era. Whereas once forestry decisions were driven solely by economics, now these decisions weigh timber supply as just one of the many values available from the forest.

Environmental groups have redefined old-growth as ancient forest, and tree farms as mono-cultural lands devoid of biodiversity. A growing public outcry for protection and preservation of forest and trees gets more attention than the demand for wood and paper products. Yet the demand for these products keeps growing.

The world population grew to 4.5 billion in 1980, and advanced to 5.3 billion by the end of the following decade. The 20th century is ending with 6.1 billion inhabiting the earth and demanding food, clothes, shelter, and other goods and services. Their cry for more goods and a voice in how the world forests are managed presents what may appear to be a mission impossible.

But the forest industry has already handled plenty of those. It successfully negotiated a rocky road, with frequent twists and turns, in the closing decades of this millennium. The world's frontier era is over; the Industrial Age has given way to the Information Age; and the forestry industry is meeting—even exceeding—the demand for forest products in new and exciting ways. What's ahead in the new century? What are the characteristics of a successful company in the new century? What is the role of the forest industry entrepreneur?

What's Ahead in the New Century

Events of the past three decades have shaped what we can expect in the new century. The forest industry is traditionally competitive, market-driven, and resilient. That competitiveness and resiliency will be sorely tested as emerging trends play out over the coming decades.

Here are the most significant major trends that affect the forest industry..

Communication—The Instant Facilitator

Advancing communication technology has stripped the earth of its secrets; instant global communication is as close as a PC keyboard. In scarcely a generation, Western Union has given way to the fax machine, the fax machine to e-mail. Rapid communication has made the earth a small planet. The flow of information provides instant recall and contact between the research center and the decision-maker, the consumer and the sales office. An information-driven technology is providing different raw materials, new manufacturing processes, and innovative wood-based products. The large-log era of forest products is ending; small logs are becoming smaller, and species types less important. Forest lands compete with farming and the generators of urban waste as a source of raw material.

Transportation and Tariffs: Cheaper, Lower, More Convenient

The years since the Cold War have provided a glimpse of what's ahead in transportation and tariffs. As tariff barriers lower, and nations use tariffs and free trade zones as economic incentives to build their economies, the free flow of goods between nations has become a flood. Land, air, and sea carriers vie for business, and shipping has never been so cheap or convenient.

An MDF customer in Turkey may purchase products from Chile, South Korea, or Portugal. A Japanese lumber buyer may purchase from Latvia, Sweden, Canada, or New Zealand. They base their choices on specifications and price, not on tariffs and political divisions. Some may say, "We have always been able to do that." But not with the ease or low cost with which we can do it now. It will be even easier in the years ahead.

Buying Groups and Grant Givers—The Drivers of Forest Policy

The impetus for environmental change and forest sustainability will come from the buying groups and the grant givers. The Earth First! and Greenpeace advocates will give way to the Home Depots and the MacArthur Foundations of the world. The World Bank and other world-class lending institutions will take an even more active role, as environmental good citizenship becomes a prerequisite to the consummation of supply contracts, credit approval, and equity acquisition.

The public, as forest products consumer, is concerned about the environmental condition of the planet. As the global population continues to rise and per capita consumption increases, this concern will only increase. Three critical issues will materially impact land and forest policy.

First, all parties must agree on definitions for such controversial words as "ecosystem," "sustainability," "biodiversity," "cumulative effects," and similar descriptive environmental words. Second, the industry must embrace the negotiation and acceptance of third-party verification/certification of environmental compliance by the forest manager. Third—and this stems directly from the first two issues—participants must "walk the talk" on forest policy. The forest industry will work with various levels and departments within governments domestic and international, overseeing and participating in the process.

The Definition and Role of Raw Material—An Ongoing Transition

When I say, "The forest has traditionally been the source of raw material for the forest industry," the operative words are "has traditionally been." Tree growing and forestry will continue to be important sources. However, the forest industry will find itself an active participant in solving the world's growing solid waste problem, as more and more post-consumer waste finds its way into lumber, board, paper, and other traditionally tree-based products. A network of collection centers will expand to meet the needs of both producer and consumer.

The industry will turn, increasingly, from trees to such lower-cost raw materials as agricultural waste, throwaway packaging and plastics, gypsum and other minerals, and other organic and inorganic materials. With a growing recognition of trees and solid wood as a repository of CO_2 gases, we will recycle aging and obsolete wood products to protect that repository, to prevent the squandering of solid waste, and to reduce raw material costs for new products. Plantations in the temperate, semitropical, and topical

A new generation is requiring new uses for the forest.

zones will come into their own as significant contributors of wood fiber through short-rotation, agriculturally intensive management.

The Consumer Will Have Greater Choice

In the past, purchasers of forest-based products based their choice on preference and appearance, negotiating price on those bases. Molding is a case in point. Assorted softwood and hardwood species, which produce clearwood with a uniform grain, once provided all the raw material. The choice narrowed down to a few desirable species—such as old-growth, fine-grained ponderosa pine (*Pinus ponderosa*) and hemlock (*Tsuga heterophylla*) and some imported hardwoods—with the only other consideration being whether or not the stock was fingerjointed.

Appearance still counts, but the growing cost of the traditional species and the negative publicity surrounding their old-growth sources are playing havoc with customer appeal and affordability. At the checkout counter, the buyer still considers performance, but wants to know whether the product comes from environmentally sustainable sources—and may suffer "sticker shock" when pricing traditional woods. New application and finishing methods, plus a growing choice between solid wood and composite-based offerings, are reducing costs and increasing choices. What can be said about molding can also be said about lap sidings, sheathing materials, and a variety of other wood products.

Buyers, blessed with a range of newer products and processes, no longer demand a particular species. Certainly, there will be the diehards who insist on a certain product—and pay the price. But for the most part, the old, simpler days of product choice are gone forever.

The Structure of the Forest Industry Is Changing

Big size, low cost, and profitability will not necessarily be the keys to success in the new century. Sectors of the forest industry such as the paper segment are so overcapitalized that satisfactory return on assets may elude them even in the best of markets. How about timber assets? Are they really as valuable as previously thought, given the emergence of a global transportation system, the urban forest, and Southern Hemisphere short-rotation plantations?

A large company stuck with billions of dollars in forest and conversion assets must field some tough questions from its owners and investors. It is difficult justifying an investment in a cyclic smokestack industry when the return on equity in recent years has averaged less than the net return of a U.S. Treasury bill. Watch for big-company growth from mergers and acquisitions—and the disappearance of more old companies.

Real industry expansion will occur in entrepreneurial ventures that produce new products from unconventional raw materials. An MDF plant consuming urban waste in Los Angeles may be far more profitable than a new paper machine in the southeast. The industry giants, seeking to diversify and to increase overall investor return, will forage for entrepreneurial businesses with improving track records. Many of these new businesses will be sited near major metropolitan areas; the size of the local market will have a real influence on plant size.

Forest Industry Vendors and Suppliers Will Shoulder More Research and Development Efforts

The industry will be more customer-driven than commodity-driven. The vast number of raw-material choices, conversion options, and product offerings will demand an expanded R&D effort. The entrepreneurial businesses won't be able to afford a sustained R&D effort, nor will the big companies reverse course and materially enlarge their R&D functions. Instead, they will rely on ideas generated by innovators within their organizations and will seek assistance from others to fully develop those ideas. The government-sponsored labs will focus on specific projects, such as reducing the solid waste stream and environmentally improving forestland as part of a larger landscape. Forest industry suppliers and equipment manufacturers will fill some of the gap in R&D efforts.

Further help will come from increased product and technology crossover between the forest industry and other industries. This crossover will provide broader markets and a more defined customer base. Costs will be justified by the increased reliance on advanced adhesive systems, electronics, and sophisticated processing lines, as well as the global nature of the industry.

Will the realigned R&D efforts suffice? Probably not. But the R&D efforts will be more focused than those in the past. Specific problems and opportunities will yield narrow solutions. Most of the theoretical/ basic research efforts will focus on the tree and its makeup.

A growing population will demand more and better housing.

The Markets Will Be Less Price-Sensitive and More Product-Driven

The forest products industry's historic demand and price behavior has been breathtaking at times. Too often, forest product prices have been on a roller-coaster ride, plummeting or rising with each impending recession or economic upturn. Particleboard provides a classic example.

Western 5/8-inch, interior underlayment sold for $58/MSM in December 1972, hit $130/MSM in December 1973, then plunged to $40/MSM by September 1974. And the 1970s weren't a rare example of cyclic markets. The mid-1990s provided another wild ride for North American lumber markets. Random Lengths, a price reporting service, published a framing lumber composite price of $200 late in 1990; little more than two years later, the same composite price reached nearly $500/MBM before plunging to under $300 by midyear, then soaring again to over $500/MSM by year-end 1993.

This erratic price behavior happens for several reasons. Producers that manufacture for a single North American economy or targeted export market leave themselves few options. Production capacity gets out of sync with available markets; limited product selection doesn't fill unforeseen needs; producers can't find a scarce raw material. Taken singly or combined, these make for a volatile market.

The forest industry is rapidly becoming an international business, with free flow of goods and services. Pricing and demand fluctuations will be tempered by the increasing raw-material options, the growing number of production technologies to fill a consumer's product need, and the varied economic performances of consuming and producing nations. Price spikes will still happen, but they won't be as sharp or last as long.

A rapid run-up in the market price for green softwood veneer on the U.S. west coast between late 1998 and early 1999 is a good example.

Within months, the price of 1/10 CD Douglas fir full sheets—a favored raw material for the regional LVL and plywood producers—climbed from under $40/MSM to nearly $60/MSM. But one buyer located a lower-cost option: a veneer plant in northern China peeling imported Russian larch (*Larix dahurica*) logs. The mill was managed by a Malaysian management team and equipped with Japanese equipment. Even factoring in the cost of overland and ocean freight, the wood was cheaper than the west coast Douglas fir veneer.

With global paper, lumber, and board markets, a need in one region can be balanced by overcapacity in another. Prices will be more stable; there will be more commonality among products and among nations. The forest products industry will see growing standardization—something already familiar to the international buyer of athletic shoes, clothes, and computers.

These are some of the major trends facing the forest industry in the new century. There are others, and more will develop as the industry moves forward into uncharted territory. One thing is certain: the coming decades will be like no other. It will be exciting; the opportunities may yet outnumber the participants.

Moving Ahead to the 21st Century: The Characterisitics of a Successful Company

The characteristics of a successful forest industry enterprise depend on how industry participants and analysts define the word "success." Some, who measure success by the consistency and predictability of cash flow and earnings, will opt for situations that provide certainty while participating as contract suppliers and/or owners of timber. In the past, it made sense to buy right and administer assets shrewdly—and it will still make sense in the future. But those who measure success by above-average return on effort and assets will look for a different set of characteristics.

Successful industry participants will think and act globally. The industry is emerging as one in which competition is international in scope even if markets or raw materials are regional in scope. Being agile, aware, and responsive, and being able to effectively deal with complexity—these are the winning characteristics that will typify the successful individual and company. That alone, though, is not enough. The industry also must recognize and deal with its past.

The past is replete with pioneers who recognized emerging trends and had the courage to take risks, break from prevailing practices, and move on. The years have passed; the business scene is different, and the issues have changed. Yet now, more than ever, the industry needs the leaders, the risk-takers, and the entrepreneurs. After three decades of unprecedented turbulence, change, and restructuring, on the brink of the 21st century the opportunities have never been greater. The task ahead is to recognize the heritage of the past, evaluate its worth for the future, assess the emerging opportunities, and boldly move ahead.

Bibliography

Chapter 1

Food and Agricultural Organization of the United Nations. *FAO Yearbook, 1988.* Rome: Food and Agricultural Organization of the United Nations, 1990.

———. *FAO Yearbook, 1996.* Rome: Food and Agricultural Organization of the United Nations, 1998.

Ficken, Robert E. *The Forested Land.* Seattle: University of Washington Press, 1987.

Hidy, Ralph W., Frank Ernest Hill and Allan Nevins. *Timber and Men, the Weyerhaeuser Story.* New York: The MacMillan Company, 1963.

Kintigh Family Limited Partnership records, 1973–1997.

Mater, Jean. *Reinventing the Forest Industry.* Wilsonville, OR: GreenTree Press, 1997.

Pease, Dave. "Where Is Lou Grant When We Need Him?" *Forest Industries* (May 1980).

Pinchot, Gifford. *A Primer of Forestry, Part II—Practical Forestry.* Washington, D.C.: Government Printing Office, 1905.

"Rethinking the Industry." *Wood Technology* (September 1997).

U.S. Bureau of the Census. *Statistical Abstract of the United States: 1997.* 117th ed. Washington, D.C., 1997.

Chapter 2

Adair, Kent T. "The Development of Global and Domestic Environmental Agenda." Briefing Paper, Stephen F. Austin State University, April 1995.

Calhoun, John M. "The Renaissance of Forest Policy." *Journal of Forestry* (December 1997).

"Comparing the Environmental Effects of Building Systems." *Wood the Renewable Resource No. 4.* Ottawa: Canadian Wood Council, 1997.

Covey, Stephen R. *Principle-Centered Leadership.* New York: Simon & Schuster, 1990.

Forest Health Science Panel. "Summary Report on Forest Health of the United States." University of Washington CINTRAFOR RE43A4, April 1997.

Gregerson, Hans et al. "Forestry for Sustainable Development: Making It Happen." *Journal of Forestry* (March 1998).

Lehman, Stan. "Brazil Notes Decline in Amazon Ruination." *The Oregonian* 27 January 1998.

Mater, Dr. Jean. *Reinventing the Forest Industry*. Wilsonville, OR: GreenTree Press, 1997.

———. "Perceptions of the Forest Products Industry." *Studies in Management and Accounting for the Forest Products Industry. Monograph Number 44*. Oregon State University, June 1997.

"Nature Conservancy in Commercial Forestry." *Southern Lumberman* (March 1998).

"North American: Last of the Sweet Days . . ." *World Wood* (July 1980).

Oliver, Chad. "Considerations of the Landscape Approach to Forest Management." Oregon Board of Forestry Tour, 24 July 1997.

Pinchot, Gifford. *A Primer of Forestry, Part II—Practical Forestry*. Washington, D.C.: Government Printing Office, 1905.

Swisher, Larry. "American Public Doesn't Support the Timber Industry." *The Register Guard,* 16 July 1997.

Toffler, Alvin and Heidi. *Creating a New Civilization: The Politics of the Third Wave.* Atlanta: Turner Publishing, 1994.

Webster, A. D. *Practical Forestry*. 4th ed. London: William Rider & Son, Ltd, 1905.

Chapter 3

"A Technology Vision and Research Agenda for America's Forest, Wood and Paper Industry." *Forest Products Journal* 46, 10 (October 1996).

Ellefson, Paul V. and Alan R. Ek. "Privately Initiated Forestry and Forest Products Research and Development: Current Status and Future Challenges." *Forest Products Journal* 46, 2 (February 1996).

Forest Products Laboratory. "A Visit to the USDA Forest Service Forest Products Laboratory Web Site." Internet: www.fpl.fs.fed.us.

"Forest Products Laboratory: Sustaining the Forest Resource." Internet: www.fpl.fs.fed.us. 2 April, 1998.

"Forest Research at a Glance." New Zealand Forest Research Limited, Rotorua, New Zealand. n.d.

Forest Research Institue of Malaysia. *FRIM Forest Research Institute of Malaysia*, 3d ed. Kuala Lumpur: 1997.

Forest Research Laboratory. *Biennial Report, 1994-1996*. Oregon State University, Corvallis, OR.

Forintek Canada Corp. Internet: www.forintec.ca. 30 April, 1996.

Johnson, L. Ward. "Research, Where Are You Now?" *Madison's Canadian Lumber Reporter* 47, 19 (May 16, 1997).

New Zealand Forest Research Institute Limited. *FRI Annual Report.* 1997. Rotorua, New Zealand.

———. Internet: www.foresteresearch.cri.nz. n.d.

———. *Forest Research, Creative Wood-based Solutions*. Rotorua, New Zealand. n.d.

Oregon State University. *Forest Research Laboratory Biennial Report 1994–1996.* Corvallis, OR.

USDA Forest Service. *Forestry Research West* (December 1997).

Wansell, Geoffrey. *Tycoon: The Life of James Goldsmith*. New York: Atheneum, 1987.

Whaley, Ross. "Key to the Future." *Forest Products Journal* 46, 3 (March 1996).

Chapter 4

Cheng, Jim and Sam Eng. "China Prepares for Board Industries Boom." *Panel World* (March 1997).

"In the Fertile Heights." *Wood Based Panels International* (August/September 1998).

International Bank for Reconstruction and Development/The World Bank. *1998 World Bank Atlas*. Communication Development Incorporated, Washington, D.C., 1998.

Johnson, Bryan T., Kim R. Holmes and Melanie Kirkpatrick. *1998 Index of Economic Freedom*. Washington, D.C.: The Heritage Foundation, 1998.

"Kafus Forms New Bio Composites Company. . . . " Internet: Quicken.com - News, http://quicken.excite.com/investments/news/story/?story=/news/stor.../a1465.htm&symbol=K. 8 October, 1998.

"L-P Claims Irish OSB Plant on Track for 1999 Profits." World Wide Wood. Internet: http://www.worldwidewood.com/revised/new/nmay98/lposb.htm. 23 May, 1998.

"Making a Million." World Wide Wood. Internet: http://www.worldwidewood.com/revised/feature/ftmar98/million.htm. 21 March, 1998

Chapter 5

Baldwin, Dick. "The Forest or the Trees." *Old Oregon*. Eugene: University of Oregon, spring 1993.

de Callejoh, Diana Propper et al. "Marketing Products from Sustainably Managed Forests: An Emerging Opportunity." In *The Business of Sustainable Forestry* by the Sustainable Forestry Working Group. Chicago: The John D. and Catherine T. MacArthur Foundation, 1998.

Food and Agricultural Organization of the United Nations. *FAO Yearbook, 1996*. Rome, 1998.

Forest Health Science Panel. *Summary Report on Forest Health of the United States*. 4 April 1997, reprinted by CINTRAFOR RE43A.

Guimier, Daniel Y. "Forestry Operations in the Next Century, A Canadian Perspective." *Journal of Forestry* 96, 6 (June 1998).

Hansen, Mark H. "Portable Technologies, Improving Forestry Field Work." *Journal of Forestry* 94, 6 (June 1996).

Harris, Richard R., Gary Nakamura and Greg Bloomstrom. "Weighing the Tradeoffs with Computerized Forest Planning." *Journal of Forestry* 95, 11 (November 1997).

Harwood, Joe. "Son of 64, Environmental Groups Consider New Strategies in Wake of Measure's Defeat." *The Register Guard,* 15 Nov. 1998.

International Paper. "Situation in Maine Sends Wake-up Call to Forest Industry." *Tree Lines* (January/February 1997).

Johnston, Jerry J., Dale R. Weigel and JC Randolph. "Satellite Remote Sensing, An Inexpensive Tool for Pine Plantation Management." *Journal of Forestry* 95, 6 (June 1997).

Mater, Catherine M. "Emerging Technologies for Sustainable Forestry." In *The Business of Sustainable Forestry* by the Sustainable Forestry Working Group. Chicago: The John D. and Catherine T. MacArthur Foundation, 1998.

Mater, Dr. Jean. *Reinventing the Forest Industry*. Wilsonville, OR: GreenTree Press, 1997.

McAlexander, James and Eric Hansen. "J Sainbury Plc and the Home Depot: Retailers' Impact on Sustainability." In *The Business of Sustainable Forestry*, by the Sustainable Forestry Working Group. Chicago: The John and Catherine T. McArthur Foundation, 1998.

McCartner, James B. et al. "Landscape Management through Integration of Existing Tools and Emerging Technologies." *Journal of Forestry* 96, 6 (June, 1998).

Millette, Thomas L., James D. Sullivan and James K. Henderson. "Evaluating Forestland Uses, A GIS-Based Model." *Journal of Forestry* 95, 9 (September 1997).

New Zealand Institute of Forestry. *Investing in Forestry.* Auckland, New Zealand, 8–9 April 1992.

Oregon Department of Forestry. "Oregon's Forests in the 21st Century: A Forum to Plan for Tomorrow." Draft. 4 January 1996.

Pinchot, Gifford. *Breaking New Ground*. Com. Ed. Washington, D.C.: Island Press, 1998.

Richenbach, Mark G. et al. "Ecosystem Management: Capturing the Concept for Woodland Owners." *Journal of Forestry* 96, 4 (April 1998).

Ripple, William J. "Historic Spatial Patterns of Old Forests in Western Oregon." *Journal of Forestry* 92, 11 (November 1994).

Smith, David Martyn. *The Practice of Silviculture*, 7th ed. New York: John Wiley & Sons, 1962.

Solberg, B. et al. *Long-term Trends and Prospects in World Supply and Demand for Wood and Implications for Sustainable Forest Management*. European Forest Institute, Joensuu, Finland, July 1996.

Stevens, William K. "New Eye on Nature: The Real Constant Is Eternal Turmoil." *The New York Times* (Science), 31 July 1990.

"Sustainable Forestry within an Industry Context." In *The Business of Sustainable Forestry* by the Sustainable Forestry Working Group. Chicago: The John D. and Catherine T. MacArthur Foundation, 1998.

World Bank Atlas, 1998. Washington, D.C., Communications Development Inc., 1998.

Chapter 6

Abdel-Gadir, A. Yassin, R. L.Krahmer and M. D. McKimmy. "Relationships Between Intra-Ring Variables in Mature Douglas-Fir Trees from Provenance Plantations." *Wood and Fiber Science* (April 1993).

Agbiotech Infosource. "An Introduction to Agriculture Biotechnology." Issue #1 (April 1993). http://www.agwest.sk.ca/insource.html.intro.html.

———. "Transgenic Plants: Extraordinary." Issue 8 (May 1994). http://www.agwest.sk.ca/insource.html/plant_extra.html.

———. "Using Genetic Fingerprints in Agriculture." Issue 18 (January1996). http://www.agwest.sk.ca/insource.html/fingerprints.html.

"Altering Lignin Concentrations." *The Forestry Source* (March 1998).

Associated Press. "Tree Farm a Boon to Forests, Mills." *The Register Guard*, 19 January 1997.

"Attendance Falls in Expobois Re-Vamp." Internet: www.worldwidewood.com/revised/feature/ftmar98/expobois.htm.

Baldwin, Dick. "A Changing Timber Resource: Genetically Creating the Forest of the Future." *Conference Proceedings*. Wood Technology Clinic and Show, Portland, OR, 25–27 March 1998.

———. "Wood Supply Changes Certain in 21st Century." *Wood Technology* (March 1997).

"Conservation Genetics of Tropical Forest Trees." Internet: http://www.bio.umb.edu/Bawa/labmain.html.

"Dendrome: Forest Tree Genome and Informatics Research Program." Internet: http://s27w007.pswfs.gov/Molelab/research.html.

DeWald, Laura E. and Mary Frances Mahalovich. "The Role of Forest Genetics in Managing Ecosystems." *Journal of Forestry* 95, 4 (April 1997).

Drlica, Karl. *Understanding DNA and Gene Cloning, A Guide for the Curious*. New York: John Wiley & Sons, 1984.

Grace, Eric S. *Biotechnology Unzipped, Promises & Realities*. Washington, D.C: Joseph Henry Press, 1997.

Lee, Jim. "Species Paulownia." *Southern Lumberman* (September 1998).

Lucas, Eric. "Giving Nature a Hand." *Horizon Air Magazine* (March 1997).

McKeand, Steve and Susumu Kurinobu. "Japanese Tree Improvement and Forest Genetics." *Journal of Forestry* 96, 4 (April 1998).

McKeand, Steve and Jan Svenson. "Sustainable Management of Genetic Resources." *Journal of Forestry* 95, 3 (March 1997).

"Mutant Pine's Abnormal Wood May Yield Environmental Benefits." *The Forestry Source* (September 1997).

Prakash, C. S. "Impact of New DNA Chip Technologies on Agbiotech Research." Internet: http://www.isb.vt.edu (September 1997).

Segovia, Lorenzo. "Getting Closer to Efficient Gene Discovery, in Silico." *Nature Biotechnology* 16 (January 1988).

Strauss, Steve. "Role of Genetic Engineering in Creating Tree Farms of the Future." Portland, OR: Wood Technology Clinic and Show, 25–27 March 1998.

Tomb, Jean-Francois. "A Panoramic View of Bacterial Transcription." *Nature Biotechnology* 16 (January 1998).

Tree Genetic Engineering Research Cooperative Annual Report 1995–96. Oregon State University Forest Research Laboratory.

Watson, W. F. "Spotlight on Forest Operations." *Forest Products Journal* 46, 7/8 (July/August 1996).

"Southern Institute of Forest Genetics." Internet: http://www.muse-usa.com.

Yin, Runsheng, Leon V. Pienaar, and Mary Ellen Aronow. "The Productivity and Profitability of Fiber Farming." *Journal of Forestry* 96, 11 (November 1998).

Chapter 7

Botting, Mike. "Waste Lines." Internet: www.worldwidewood.com/revised/feature/ftmar98/waste.htm. March 1998.

Center for Wood Utilization Research Progress Report for FY 1996. Forest Research Laboratory. Corvallis, OR: Oregon State University, 1996.

Dalen, Herlof. "Limitation Factors in Producing Straw Board." *Panel World* (November 1998).

Douglas, Todd. "Prairie Forest Products Appears to Have Turned the Wheat Straw Corner." *Panel World* (November 1998).

"Finding Unique Wood in a Cookie-Cutter World." *FDM* (August 1998).

"From Recycled Wood to Prime Quality Boards." *Wood Based Panels International* (April/May 1998).

Fuller, Brian. "Manager's Corner." *Bring Recycling's Used News* 7, 3 (fall 1998).

"Isobord Makes the Grade." *Wood Based Panels International* (April/May 1998).

"It's All a Matter of Straw." *Wood Based Panels International* (April/May 1998).

Johnson, Kathryn. "Fine Lumber's 'Ancient' Timber." *Building Material Retailer* (June 1998).

"Kafus Enters Fiber Cement Siding Business through Acquisition of Joint Venture with Temple Inland." Internet: http://biz.yahoo.com/bw/980403/kafus_envi_1.html. 3 April, 1998.

Kafus Environmental Industries Corporate Profile. Internet: www3.stockgroup.com/kafu/index1.html.

Louisiana Pacific Corp. "Nature Guard Insulation." October 1994.

Lyman, Mark. "Profitable Opportunities Exist in Wood Waste Processing." *Modern Woodworking* (September 1998).

National Wood Recycling Directory. American Forest and Paper Association (January 1996).

"Producing Pulp on Agricultural Lands." *The Forestry Source* (June 1998).

Suchsland, Otto, et al. "Laboratory Experiments on the Use of Recycled Newsprint in Wood Composites." *Forest Products Journal* 48, 11/12 (November/December 1998).

"The News." *Wood Technology* (May 1998).

"The News." *Wood Technology* (June 1998).

"The Pied Piper of Sun Valley." Internet: www.worldwidewood.com/revised/report/tcjan99.inorgan.htm. January 1999.

U.S. Bureau of the Census. *Statististical Abstract of the United States: 1997* (117th edition.) Washington, D.C., 1997.

USDA Forest Service. *Housing: A Home for Recycled Materials.* March 1994.

Williams, Ward. "A Balance with Nature." *Wood Based Panels International* (August/September 1998).

Wood Technology, Annual International Panel Review (October 1998).

Chapter 8

American Plywood Association Management Report 4 (October 1991).

APA-The Engineered Wood Association. "Engineered Wood Product Demand and Production Continue on Record-Setting Pace." Internet: http://biz.yahoo.com/prnews/980310/wa_apa_woo_1.html. 10 March, 1998.

———. "Engineered Wood Statistics, First Quarter 1998." Tacoma: May 1998.

———. "PRP-108, Performance Standards and Policies for Structural-Use Panels." February 1994.

———. "PS1-95, Construction and Indurstrial Plywood". September 1995.

———. "PS2-92, Performance Standard for Wood-based Structural-Use Panels." August 1992.

Baldwin, Richard F. *Plywood and Veneer-Based Products: Manufacturing Practices.* San Francisco: Miller Freeman Publications, 1995.

Barr, Linda Keller. *The Engineered Wood Solution.* Portland: C. C. Crow Publications, 1995.

"Building With Trex." Trex Company, LLC, 1997.

Canadian Standard CSA 0325 Trademarking Guidelines. January 1993.

Casillo, Romulo C. and Sue-zone Chow. "Press-Time Reduction by Preheating and Strength Improvment by Finger-Jointing Laminated Veneer Lumber." *Forest Products Journal* 29, 11 (November 1979).

Donnell, Rich. "LP Case Will Go Down as One of Industry's Saddest." *Panel World* (July 1998).

Engineered for success. Internet: http://www.worldwidewood.com/revised/feature/nasoft/engine.htm. 25 April, 1998.

Japan Plywood Inspection Corporation. *Japanese Agricultural Standard for Structural Laminated Veneer Lumber.* Tokyo, 14 September 1988.

Koenigshof, Gerald A. "Status of COM-PLY Floor Joist Research." *Forest Products Journal* 29, 11 (November 1979).

Kurpeil, Frederick. *The Attributes of a Still-Evolving Product—Plywood.* Report prepared for R. F. (Dick) Baldwin. 1994.

Lorenz, Ray. "How They View Engineered Wood Quality." *Building Material Retailer* 14, 4 (August 1996).

Louisiana Pacific Corporation Annual Report 1996.

Lumber and Panel Markets Through 1986 Forsim Review. Data Resources, Inc. December 1983.

Maloney, Thomas M., ed. *Proceedings 29th International Particleboard/Composite Materials Symposium W.S.U. 1995*. Pullman, WA: Washington State University, 1995.

"MDF Reaps the Rewards of 'Imaginative Promotion.'" Internet: http://www.worldwidewood.com/revised/market/mkjun98/ireland.htm. 6 June, 1998.

Rasback, Roger. *The Provident Home*. Houston: Provident Press, 1993.

Schumacher, Rick. "Engineered Wood Market Continues to Grow." *Building Material Retailer* (August 1998).

Shupe, Todd F., et al. "Effect of Silvicultural Practice and Veneer Grade Layup on Some Mechanical Properties of Loblolly Pine LVL." *Forest Products Journal* 47, 9 (September 1997).

Smulski, Stephen, ed. *Engineered Wood Products, A Guide for Specifiers, Designers and Users*. Madison: PFS Research Foundation, 1997.

Sweden Plans Innovative "Star-Cutting" Facililty. Internet: http://www.worldwidewood.com/revised/news/njun98/starcut.htm. June 1998.

Terms of the Trade. Eugene, OR: Random Lengths Publications, 1993.

"TJ International Reports Increase Sales and Earnings in the Second Quarter." Internet: http://quicken.excite.com/investments/n...ory=/news/stories/pr/19980720/a1950.htm. 20 July, 1988.

"Triboard: Offering the Best of Both Worlds?" Internet: http://www.worldwidewood.com/revised/reprot/nzealand/triboard.htm. April/May 1998.

UDM Upholstery Design and Manufacturing 11, 8 (September 1998).

Williamson, Thomas G. "A PA-EWS PRI-400, A Performance Standard for I-Joists." *PETE Conference Proceedings, Markets and Manufacturing in Panels/Composites for '97*. Atlanta: Wood Technology, 1997.

Wood Products Review. Resource Information Systems, Inc. March 1996.

Wood Reference Handbook. Ottowa: Canadian Wood Council, 1991.

Chapter 9

Adhesive Bonding of Wood. Technical Bulletin No. 1512. USDA Forest Service. August 1975.

Atchison, Darlene, et al. "Ecological & Economic Significance of Glue Extenders & Fillers in Plywood Production." *1997 Plywood Manufacturing Short Course*. Oregon State University, Corvallis OR, 23 April 1997.

"Award-Winning Innovation: Reinforced Glue-Laminated Timber." *Biennial Report, 1994–1996*. Forest Research Laboratory, Madison, WI.

Baldwin, Richard F. *Plywood and Veneer-Based Products Manufacturing Products*. San Francisco: Miller Freeman, 1995.

Chow, Sue-zone, and Paul R. Steiner. "Comparison of Curing and Bonding Properties of Particleboard- and Waferboard-Type Phenolic Resins." *Forest Products Journal* 29, 11 (November 1979).

Conner, Anthony H., et al, eds. *Wood Adhesives 1990.* Forest Products Laboratory, Madison, WI, 1990.

Erickson, John R. "The Role of Adhesives in the Improved Use of Our Timber Resources." Keynote Address, 4th Annual International Symposium on Adhesion and Adhesives for Structural Materials. Washington State University, October 1984.

"Finger Jointing with Soy-Based Adhesives." *Modern Woodworking* 12, 6 (June 1998).

"First Mill Certified to Use Soy Adhesive." *Feedstocks* 2, 6 (January 1998).

Gillespie, Robert H., et al., eds. *Adhesives in Building Construction.* USDA Agriculture Handbook No. 516 (February 1978).

Haupt, Robert A., and Terry Sellers Jr. "Phenolic Resin-Wood Interaction." *Forest Products Journal* 44, 2 (February 1994).

Hse, Chung-Yun, et al., eds. *Adhesives and Bonded Wood Products.* Proceedings No. 4735. Madison, WI: Forest Products Society, 1991.

"IMAL Glueing Technology for Particleboard Plants." Advertisement, *Wood Based Panels International* (August/September 1998).

Marra, Alan A. *Technology of Wood Bonding: Principles in Practice.* New York: Van Nostrand Reinhold, 1992.

Massey, Rick. "Problem: Searching for a Value-Added Alternative to Basic Substrate Panels. Solution: Using Overlays." *FDM* (September 1998).

McLaughlin, Alexander, et al. "Polymeric Isocyanate as Wood Product Binders." *Proceeding of 1980 Symposium "Wood Adhesives—Research, Applications and Need."* Forest Products Laboratory, Madison, WI, 1980.

Pizzi, A. *Advanced Wood Adhesives Technology.* New York: Marcel Dekker, Inc., 1994.

Sellers Jr., Terry. "Adhesive Industry Matching Wood Composite Needs." *Panel World* (September 1998).

———. "Gluing of Eastern Hardwoods: A Review." *USDA General Technical Report SO-71* (September 1988).

———. "Overview of Wood Adhesives in North America." Presentation, 52nd Annual meeting, Forest Products Society, Merida, Yucatan, Mexico, 21–24 June, 1998.

Sellers Jr., Terry, and Robert A. Haupt. "New Development in Wood Adhesives and Gluing—Processes in North America." *Crows*, October 1995, December 1995, April 1996.

Simonds, John E., et al. "Estimating Cogeneration Feasibility: A Computer Model." *Forest Products Journal* 42, 9 (September 1992).

"Simpson Select." Simpson Timber Company. Portland, OR. n.d.

"Soy-Based Foaming Adhesive May Change Plywood Industry." *Feedstocks* 3, 1(March 1998).

"Soy-Enhanced Extender Products Boost Performance of Plywood Adhesives." *Feedstocks* 3, 2 (May 1998).

White, James T. "Growing Dependency of Wood Products on Adhesives and Other Chemicals." *Forest Products Journal* 29, 11 (November 1979).

Chapter 10

Crook, Colin. "Complexity and Technology: The Business Perspective." *Conference Proceedings, Complexity and Strategy*. Santa Fe Institute, London, 15–17 May 1995.

de Geus, Arie. "Organizational Principles of Long-Term Corporate Survivors." *Conference Proceedings, Compexity and Strategy*. Santa Fe Institute, London, 15–17 May 1995.

"Examine your Corporate Strategy." *Forintek Review II/I.*

Labich, Kenneth. "Why Companies Fail." *Fortune* 14 (November 1994).

Rich, Stuart U. "Meeting the Challenge of Change." 13th Annual Portland Business Conference, Portland OR, 17 February 1971.

Wansell, Geoffrey. *Tycoon: The Life of James Goldsmith*. New York: Atheneum, 1987.

Chapter 11

Aeppel, Timothy. "Why 'Too Much Stuff' Means Little Now." *Wall Street Journal*, 4 December 1997.

Baldwin, Richard F. "An Accountant as an Advisor." *Personal Notes*, 2 February 1987.

Cirillo, Mary A. "Techniques for Management in Transition in Complex Organizations." *Conference Proceedings, Complexity and Strategy*. Santa Fe Institute, London, 15–17 May 1995.

Crook, Colin. "Complexity and Technology: The Business Perspective." *Conference Proceedings, Complexity and Strategy*. Santa Fe Institute, London, 15–17 May 1995.

Davis, L. J. "They Call Him Neutron." *Business Month* (March 1988).

Finegan, Jay. "Against the Grain." *Inc.* (November 1992).

Kaufman, Stuart, and William Macready. "Technological Evolution and Adaptive Organizations." *Conference Proceedings, Complexity and Strategy*. Santa Fe Institute, London, 15–17 May 1995.

Petzinger Jr., Thomas. "The Front Lines: A Plant Manager Keeps Reinventing His Product Line." *The Wall Street Journal*,19 September 1997.

Stern, William. "Of Mules and Men." *Forbes*, 8 November 1993.

Chapter 12

Baldwin, Dick. "Wood Supply Changes Certain in 21st Century." *Wood Technology.* (March 1997): 28-39.

Business of Sustainable Forestry: Case Studies. Sustainable Forestry Working Group. MacArthur Foundation, 1998.

Food and Agricultural Organization of the United Nations. *FAO Yearbook*. Rome: Food and Agricultural Organization of the United Nations. 1995 ed., 1994 ed., 1998 ed.

Johnson, Kathryn. "Fine Lumber's 'Ancient' Timber." *Building Material Retailer* (June 1998): 22–23.

"Oregon's 1994 Timber Harvest Hits Another Record Low." *Forest Log*. (Oregon Department of Forestry, Sept.–Oct. 1995)

Porterfield, Richard L., and John B. Crist, eds. "Impacts of the Changing Quality of Timber Resources." *Proceedings of the Forest Products Research Society*. Atlanta, June 1978: 22.

Statistical Abstract of the United States: The National Data Book. US Department of Commerce, Economics and Statistics Administration, Bureau of the Census, 1997.

Chapter 13

1996 Annual Report. TJ International, Boise ,ID.

1997 Annual Report. TJ International, Boise ID.

Annual Financial Report 1985. Louisiana Pacific, Portland, OR.

Annual Report 1995. Louisiana Pacific Corporation, Portland, OR.

Annual Report 1996. Louisiana Pacific Corporation, Portland, OR.

"Examine Your Corporate Strategy." *Forintec Review II/I*.

Forbes, Craig L., Sarah Griffin Peck, and E. T. Altman. *Customer Oriented Marketing for the Hardwood Plywood Industry*. Hardwood Plywood and Veneer Association (fall 1996).

Guss, Leonard M. "Engineered Wood Products: The Future Is Bright." *Forest Products Journal* 45, 7/8 (July /August 1995).

Johnson, Angela. "The Dealer/Distributor Partnership." *Building Material Retailer* (June 1998).

Kozak, Robert A. "How Specifiers Learn about Structural Materials: A Study in Value-Added Marketing." *Preparing for the 21st Century: Value Added Marketing for Value Added Wood Products*. IUFRO and FPS Conferences, June 1997.

"MacBlo Slumps into Heavy Loss but Cedar and Composites Remain Strong." Internet: www.worldwidewood.com. 21 February, 1998.

MacMillan Bloedel, Internet. www.mbltd.com. April 1998.

Mater, Dr. Jean. "'Value-Added' Products: Key to Competition." *Forest Industries* (March 1990).

"MB Wants to Sell Two MDF Plants." *Panel World*. (March 1998).

Plywood Pioneers Association. "Washington Veneer Company." Monograph #11, Tacoma, September 1971.

Sherrill, Dr. Sam. Personal conversation with author, January 1999.

Thieles, Bart A. Personal letter to author, 12 August 1996.

Thompson, Jay. "MacBlo Eyes Euro Markets after Massive Restructuring," Internet: www.worldwidewood.com. 31 Jan, 1998.

US Bureau of the Census. *Statistical Abstract of the United States: 1993*. (113th ed.) Washington, D.C., 1993.

———. *Statistical Abstract of the United States: 1996*. (116th ed.) Washington, D.C., 1996.

Vlosky, Richard P., et al. "Partnerships Versus Typical Relationships between Wood Products Distributors and Their Manufacturer Suppliers." *Forest Products Journal* 48, 3 (March 1998).

Whaley, Ross S. "Technological Adaptation: The Key to Our Future." *Forest Products Journal* 46, 3 (March 1996).

Chapter 14

Accord, Terry. "Flow Works for Vendors, Too." *FDM* (April 1998).

Adams, Larry. "Cabinet Industry Continues Growth Surge." *Wood & Wood Products* (April 1998).

"Award-Winning Innovation: Reinforced Glue-Laminated Timber." *Biennial Report 1994–1996*, Forest Products Laboratory.

Blackman, Ted. "Lots of Lumber, Lower Lemand Forecast a WWPA Spring Meet." *Wood Technology* (May 1998).

Fossom, Harold. "Secondary Products Hold Key to Future of Woods" *Register Guard* 21 (December 1997).

Gohmann, C. L. Letter to author. 25 April, 1998.

Harwood, Joe. "Lumber Mill May Stay Shut," *Register Guard* 8 (January 1998).

Kurpeil, Frederick T. "Veneer and Composite Based Panels and Engineered Lumber." *Conference Proceedings*. Wood Technology Clinic & Show, Portland, OR, March 1998.

Lawser, Steven V. "Secondary Wood Processing Overview." *Conference Proceedings*. Wood Technology Clinic & Show, Portland, OR, March 1998.

Lettman, Gary. "Timber Availability from Nonfederal Forest Land in Oregon." *Conference Proceedings*. Wood Technology Clinic & Show, Portland, OR, March 1998.

"Lumber, Panel Products Looking for Direction." *Crows* (21 June 1996).

"Making the Grade." Internet: www.worldwidewood.co.uk. 7 February, 1998.

Mater, Dr. Jean. "Effective Niche Marketing." *Conference Proceedings*. Wood Technology Clinic & Show, Portland, OR, March 1998.

Mater, Dr. Jean. "'Value-Added' Products: Key to Competition." *Forest Industries* (March 1990).

Michael, Judd H. "Tactics for Enticing Value-Added Wood Manufacturers." *Texas Forestry* (December 1997).

Pease, Dave. "Measuring the Value of Value." *Wood Technology* (May 1998).

Peters, Thomas J., and Robert H. Waterman Jr. *In Search of Excellence*. New York: Harper & Row, 1982.

Peterson, Wade A. Letter to Wallace Glausi. 31 March 1998.

"Quotas? Canadians Bypass Them by 'Adding Value.'" *Wood Technology* (May 1998).

Susnis, Chuch. "Shedding Some Light on UV Curable Coatings." *Wood & Wood Products* (April 1998).

Chapter 15

Adair, Kent. Personal letter to author. 17 January 1996.

Banzhaf, William. "Forest Certification: Diversity of Opinion Ensures Engagement." *Forestry Source* (March 1998).

Barbour, James R., and Kenneth E. Skog, eds. *Role of Wood Production in Ecosvstem Management*. Proceedings of the Sustainable Forestry Working Group at the IUFRO All Division 5 Conference, Pullman, WA, July 1997. USDA FPL-GTR-100.

Berg, Scott. "Book Review: Certification of Forest Products: Issues and Perspectives." *Forest Products Journal* (April 1997).

Berg, Scott, and Julie Jack. "Forest Products Industry Voluntarily Ensures Environmental Quality." *Forestry Source* (September 1997).

"Canada's First Certification Goes to FSC Via SmartWood." Internet: www.worldwide-wood.co. 21 March, 1998.

Carter, Douglas R., and Frank D. Merry. "The Nature and Status of Certification in the United States." *Forest Products Journal* (February 1998).

"Deforestation up." Internet: www.worldwidewood.com. 31 January, 1998.

Dixon, Audrey. "Developed Countries Are Failing Poorer Nations, Says Premier." Internet: www.worldwidewood.com. 14 March, 1998.

———. "World Bank Reveals Some Internal Divisions over Forest Certification." Internet: www.worldwidewood.corn. 14 March, 1998.

Domtar. *Annual Report*. Quebec, 1997.

Donovan, Richard. "Certification Has Brought on the Age of the Independent Forest Auditor." *Forestry Source* (September 1997).

Fialka, John J. "Breathing Easy: Clear Skies Are Goal as Pollution Is Turned into a Commodity." *Wall Street Journal*, 3 October 1997.

Forest Stewardship Council. "United States Initiative." Waterbury, VT. n.d.

"'Green' Bottom Line Looks Good." Internet: www.worldwidewood.com. 7 March, 1998.

"Greenpeace Calls for BC Rainforest Boycott." Internet: www.worldwidewood.com. 7 March, 1998.

"Halifax at Home in US Hardwoods." Internet: www.worldwidewood.com. 21 March, 1998.

Hammel, Debbie. Letter to Tom Clouts. 6 March 1996.

Hansen, Eric. "Forest Certification and Its Role in Marketing Strategy." *Forest Products Journal* (March 1997).

Hartzell, Gate, and Steve Gretzinger. "Public Lands Certification." *Community Ecology*. Rogue Institute for Ecology and Economy. Ashland, OR (summer/fall 1997).

Hupe, Kurt. "Timber Certification: Fact or Fad?" *The Merchant Magazine* (May 1997).

Mater, Dr. Jean. "Perceptions of the Forest Products Industry." Monograph Number 44. Oregon State University, June 1997.

Moore, Henson. "Reinventing the Forest Products Industry." Technical Paper 96-P-16. American Pulpwood Association, 16 April 1996.

Moore, Patrick. "Green Bans Won't Save the Forests." *Canberra Times* (Australia), 14 July 1997.

"Newly Formed Public Interest Committee Has Big Name, Big Job." Oregon Dept. of Forestry News Release, 13 April 1998.

"Northern California SAF Addresses Certification." *The Forestry Source* (March 1998).

Ozanne, Lucie K., and Richard P. Vlosky. "Willingness to Pay for Environmentally Certified Wood Products: A Consumer Perspective." *Forest Products Journal* (June 1997).

Pease, Dave. "Wood Industry Outlook Steady through 1998." *Wood Technology* (January/February 1998).

"Price and Pierce Backs Certified Brazilian MDF and Hardboard." Internet: www.worldwidewood.com. 7 March, 1998.

Ruddell, Steve, and James A. Stevens. "The Adoption of ISO 9000, ISO 14001, and the Demand for Certified Wood Products in the Business and Institutional Furniture Industry." *Forest Products Journal* (March 1998).

"SAF Established Certification Task Force." *Forestry Source* (April 1998).

"SmartWood Certified Forestry." Rogue Institute for Ecology and Economy. Ashland, OR. Undated.

"Supreme Court to Hear Ohio Case this Month." *Southern Lumberman* (February 1998).

Strategic Planning 1997–1999. Oregon Department of Forestry.

"Swedish Forest Products to Have Eco-Labeling." *Wood Technology* (January/February 1998).

Viana, Virgilio M., et al, eds. *Certification of Forest Products: Issues and Perspectives*. Washington, D.C.: Island Press, 1996.

Williamson, Thomas G. "APA-EWS PRI-400: A Performance Standard for I-Joists." *Conference Proceedings*, PETE, 20–23 October 1997.

Wilson, Alex, and Nadav Malin. "Wood Products Certification: A Progress Report." *Environmental Building News* (November 1997).

Winistorfer, Steve G., and Harold J. Steudel. "ISO 9000: Issues for the Structural Composite Lumber Industry." *Forest Products Journal* (January 1997).

Chapter 16

Baldwin, Catherine A. *Making the Most of the Best: Willamette Industries' Seventy-Five Years*. Portland, OR: Willamette Industries, 1982.

Beatty, Jack. *The World According to Peter Drucker*. New York: The Free Press, Simon & Schuster, 1998.

Correll, A. D. "Growing Toward a Sustainable Future." *Technical Paper 96-P-15*. American Pulpwood Association, Inc., 15 April 1996.

Covey, Stephen R. *Principle-Centered Leadership*. New York: Fireside, Simon & Schuster, 1990.

———. *The 7 Habits of Highly Effective People*. New York: Simon and Schuster, 1989.

Covey, Stephen R., Roger A. Merrill, and Rebecca R. Merrill. *First Things First*. New York: Fireside, Simon & Schuster, 1994.

Cox, H. J. *Random Lengths: Forty years with 'Timber Beasts" and "Sawdust Savages."* Eugene, OR: H. J. Cox, 1949.

Drucker, Peter. *Innovation and Entrepreneurship: Practice and Principles*. New York: Harper & Row, 1985.

"Kimberly-Clark Forms Global Strategic Leadership Teams." Kimberly-Clark Corp. press release, 10 December 1998.

Mater, Dr. Jean. *Studies in Management and Accounting for the Forest Products Industry: Perceptions of the Forest Products Industry, Monograph no. 44*. Oregon State University, June 1997.

"Weyerhaeuser Announces Organizational Changes to Improve Manufacturing and Customer Focus." Weyerhaeuser Company press release, 7 December1998.

Chapter 17

Adair, Kent T. "A Better Way: Save the Planet with Free People Still on It!" *Texas Forestry* (April 1997).

———. "The Development of Global and Domestic Environmental Agenda." Briefing Paper, Steven F. Austin State University, April 1995.

American Forest Service. *Project learning Tree Environmental Education Activity Guide*. Washington, D.C., 1995.

Benson, Mitchel. "Timber Industry Is Trying to Paint Itself Green." *Wall Street Journal*, 22 April 1998.

"Canada Calls for Global Forestry Convention." *Forestry Source* (June 1997).

"Champion International Corporation to Voluntarily Open All U.S. Forest Operations to Third-Party Reviews in Support of the Sustainable Forestry Initiative Program." Internet: http://quicken.excite.com/investments/news/story/?story=/news/st.../a1848.htm&symbol=CH. Nov. 16, 1998.

"Clear-Cutting Is Still an Issue after Maine Referendums Fail." *New York Times*, 12 November 1997.

Currie, Bob. "Texas Squeezed by Global Problems." *Texas Forestry* (July 1998).

Contreras-Hermosilla, Arnoldo. "The 'Cut-and-Run' Course of Corruption in the Forestry Sector." *Journal of Forestry* 95, 12 (December 1997).

Crowell, David. "They Don't Cut and Run." *Register Guard*, 21 January 1998.

"Deforestation up." Internet: www.worldwidewood.co.uk/revised/news/njan98/deforest.htm. 31 January, 1998.

Diener, Betty J. "Portico, S.A.: Strategic Decisions 1982–1997." *The Business of Sustainable Forestry*. Chicago: The John D. & Catherine T. MacArthur Foundation, 1998.

"Environment-Chile: New Measures Against Logging Project." Inter Press Service, 4 June 1998.

Ferdinand, Pamela. "Conservationists Buy Maine Woodland." *Register Guard*, 16 December 1998.

Geisinger, Jim. "Clinton's Forest Plan Turned Woods into a Money Loser." *Register Guard*, 23 Oct., 1997.

"Globalization: Surfing the FiberNet." *Crow's* 11, 6 (December 1996).

Gregersen, Hans, Allen Lundgren and Neil Byron. "Forestry for Sustainable Development: Making It Happen." *Journal of Forestry* 96, 3 (March 1998).

Helm, Mike. "Cost of Logging Is High." *Register Guard*, 21 January 1998.

Jukofsky, Diane. "Deforestation Also to Blame for Damage Left by Mitch." *Register Guard*, 25 November 1998.

Kanamine, Linda. "Northwest May Face Sweeping Changes." *USA Today*, 4 April 1997.

Kleine, Ted. "Libertyville Plans to Restore Cornfield to Prairie." *Chicago Tribune*, 26 October 1997.

Legg, Bob. Personal correspondence with author. 8 January, 1999.

Lehman, Stan. "Brazil Notes Decline in Amazon Ruination." *Oregonian*, 27 January 1998.

Lichtman, Pamela. "The Politics of Wildfire: Lessons from Yellowstone." *Journal of Forestry* 96, 5 (May 1998).

"MacMillan Bloedel to Phase Out Clearcutting, Old-Growth Conservation Is Key Goal, Customers to Be Offered Certified Products." Internet: http://quicken.excite.com/investments/n...ory=/news/stories/pr/19980610/a3018.htm. 10 June, 1998.

"MacMillan Bloedel to Stop Clearcutting." *Forestry Source* 3, 7 (July/August 1998).

Masterson, Lance. "Fremont Sawmill Earns Auditors Highest Rating." *Lake County Examiner*, 30 April 1998.

Mater, Dr. Jean. *Studies in Management and Accounting for the Forest Products Industry: Perceptions of the Forest Products Industry, Monograph no. 44*. Oregon State University, June 1997.

Nelson, Deborah, et al. "Trading Away the West: How the Public Is Losing Trees, Land and Money." *Seattle Times*, 27 and 28 September 1998.

Nesmith, Jeff. "Poll: Laws Not Green Enough, Most Feel." *Register Guard*, 5 June 1998.

Paquette, Michael J. "Church Council Joins Global Warming Fight." *Oregonian*, 20 August 1998.

"Restoring Lumber's Mildewed Image." *National Home Center News*, 23 November 1998.

"Rockers vs. Renewable Resources?" *Building Material Dealer* (October 1998).

Saffell, William H. Letter to editor. *Journal of Forestry* 96, 8 (August 1998).

Shindler, Bruce, and Julie Neburka. "Public Participation in Forest Planning: 8 Attributes of Success." *Journal of Forestry* 95, 1 (January 1997).

Suzuki, Yumiko. "National Forests Teeter on Edge of Crisis." *Nikkei Weekly*, 16 June 1997.

Swisher, Larry. "American Public Doesn't Support Timber Industry." *Register Guard*, 16 July, 1997.

"Timber Plan Will Protect Wildlife." *Register Guard*, January 31 1997.

Warrick, Judy. "A Warning of Mass Extinction." *San Francisco Chronicle*, 21 April 1998.

Chapter 18

Bilek, Edward M., and Paul V. Ellefson. "Stock Ownership of Selected Public U.S. Wood-Based Corporations." *Forest Products Journal* 36, 3 (March 1986).

Brinkman, Jonathan. "Crown Pacific Plays by Its Own Rules and Wins." *Oregonian*, 21 June 1998.

Drucker, Peter F. *Managing in Turbulent Times*. New York: Harper & Row, 1980.

———. "A Turnaround Primer." *Wall Street Journal*, 2 February, 1993.

Ehinger, Paul F., and Robert Flynn. *Forest Products Industry: Report on Mill Closures, Operations, and Other Related Information*. Eugene, OR: Paul Ehinger and Associates. December 1993.

Kawasaki, Guy. "How to Drive Your Competitors Crazy." *Wall Street Journal*, 17 July, 1995.

McMaster, Michael D. *The Intelligence Advantage, Organizing for Complexity*. Boston: Butterworth-Heinemann, 1996.

"The Technology That Is Nudging Plywood Aside." *Business Week* , 28 May, 1984.

Walton, Sam. *Sam Walton Made in America: My Story*. New York: Doubleday, 1992.

Chapter 19

Hebblethwaite, Erika. "A Sound Proposal." Internet: www.worldwidewood.com/revised/feature/nasoft/clay.htm. 25 April, 1998.

"Nature Conservancy in Commercial Forestry." *Southern Lumberman* (March 1998).

Robinson, Gordon. *The Forest and the Trees*. Washington, D.C.: Island Press, 1988.

Toffler, Alvin and Heidi. *Creating a New Civilization: The Politics of the Third Wave*. Altanta: Turner Publishing Co., 1994.

Glossary

Evans, David S., ed. *Terms of the Trade*, 3rd ed. Eugene, OR: Random Lengths Publications,1993.

Marra, Alan A. *Technology of Wood Bonding, Principles in Practice*. New York: Van Nostrand Reinhold, 1992.

Robinson, Gordon. *The Forest and the Trees, A Guide to Excellent Forestry*. Washington D.C.: Island Press, 1988.

Glossary

$3/8$ths basis—Often expressed as M $3/8$ths. The volume measurement standard of the North American structural panel and softwood veneer industry. Based on a tally of square footage, equivalent (real or nominal) $3/8$ths inch thick.

6-ply $23/32$ AC—Plywood that is made with 6 plies, $23/32$ inches thick, real, or nominal, with an A-grade face, C-grade back, and C-grade inner plies. Used in applications where only one side needs a smooth appearance.

A-grade veneer—Veneer with a smooth-paint-grade face that has limited repairs.

American Forest & Paper Association (AF&PA)—An industry group of the North American forest industry.

American Lumber Standards (ALS) Committee—An association that promotes and accredits organizations that monitor compliance to a softwood lumber product standard.

Ancient timber—Old lumber recovered from a structure being remodeled or dismantled. Also known as urban lumber.

Billet—The rough, hot-pressed laminated veneer assembly prior to being remanufactured into smaller sizes. (Also see Laminated Veneer Lumber)

Board foot—A unit of measurement, used in North America for logs and lumber, that represents equivalent square-footage surface measure by real or nominal 1-inch thickness.

BM—Board measure.

Buying groups—A cooperative of sellers who purchase for all members of the cooperative. The objectives of the group are many and varied: using collective buying power to purchase cheaper is one; obtaining cooperative objectives, such as obtaining certified wood, is another.

Check—A wood fiber separation along the length of the tree, log, veneer, lumber, or other wood component. Usually caused by natural or imposed drying stress.

Common—1. A lumber term applied to board sizes. 2. Lumber that is suitable for general construction and utility purposes.

Crossband—A veneer or composite ply within plywood, LVL, and various composite panels that has the wood grain positioned perpendicular to the face and back ply. Sometimes called core or x-band.

Curtain-coated—A method of glue application in which the veneer passes through a "curtain" of adhesive.

Cut-over—Land that has been previously logged

D graded—A softwood veneer grade that is usually designated as the back-ply or inner-ply for rough sheathing or other plywood panels that have limited exposure to moisture in service.

Diameter Breast Height (DBH)—A commonly used point of measurement in estimating the wood volume in a standing tree. Generally 4.5 feet, or 1.37 meters, above ground level.

Doyle log rule basis—One of the many log rules used to determine the number of board feet in a log. The Doyle log rule is most commonly used in the southeastern United States.

Eco-culture—Management of the entire forest ecosystem.

Eco-labeling—Labeling a product stating that it comes from an ecologically sustainable resource.

Electronic data recorder—An electronic device for recording data.

Electronic distance measurement device—An electronic device for measuring distance.

Engineered Strand Lumber (ESL)—A proprietary lumber product created with strands of fiber that are layered lengthwise and bonded together. Specifically designed for use in upholstered furniture frames as a replacement for both lumber and plywood.

Engineered Structural Panels (ESP)—Panel products designed to meet particular specifications concerning strength, rigidity, etc. for structural use.

Engineered Wood Products (EWP)—Lumber or panels manufactured by using adhesives to hold together oriented veneers, wafers, wood fibers, or dimension lumber.

Feller-buncher—A tree-cutting machine that extends hydraulically activated arms around the tree trunk and severs the tree at the base. The feller-buncher then pivots with the tree stems and stacks neat piles of harvested trees

Fingerjointing—A method of joining two pieces of lumber end-to-end by sawing into the end of each piece a set of projecting "fingers" that interlock. When the pieces are pushed together and glued, a strong joint is formed.

***Fomes pini* rot, or lacey white spec**—A fungus that develops in an overmature tree that causes clusters of small white areas in the heartwood. Also called white pocket.

Forest Engineering Research Institute of Canada (FERIC)—An association whose purpose is to improve Canadian forestry operations related to the harvesting and transportation of wood and the growing of trees within a framework of sustainable development.

The Forest Stewardship Council (FSC)—A group that accredits the certifiers of environmentally sustainable forests.

Front-end loader—A machine that has log-loading apparatus on the front so the operator can watch as he moves the log.

Global positioning system (GPS)—An electronic device that determines geographical position.

Glulam—Also called laminated beams. (See Lamstock.) A process in which individual pieces of lumber or veneer are bonded together with an adhesive to make a single piece, with the grain of each piece running parallel to the grain of each of the other pieces.

Gmelina—(*Gmelina arborea*) A fast-growing tropical tree commonly grown as a plantation species in lowland tropical areas.

Greenweld process—A New Zealand innovation that uses a melamine-urea resin and a hardener containing resorcinol in combination with a curing agent. This allows the fingerjointing of green lumber, usually short lengths, into specified lengths.

High-pressure laminate—A sheet of material formed from multiple layers of kraft paper saturated with phenolic resin, a decorative layer of paper saturated with melamine resin, and a thin top sheet of paper saturated with melamine resin. The layers are pressed together under high heat and pressure to form a stiff plastic sheet.

I beam—A structural beam composed of a top and bottom flange, with a plywood or OSB webbing material.

I-joists—A parallel joist used to support floor and ceiling loads, made by using adhesive to attach wood flanges to a plywood or OSB web.

International Organization for Standardization (ISO)—An international standards body that has developed a system-based performance standard, including incorporating environmental (14000 series) and quality (9000 series) objectives and targets into an overall management system.

Intrallam—Proprietary name for OSL/LSL in Europe.

ISO 9000 protocol—The rules of the ISO standard for quality management. (See International Organization for Standardization.)

Kiln-dried (KD) lumber—Lumber that has been seasoned in a kiln to a predetermined moisture content.

Lacey white spec, or *fomes pini* rot—A fungus that develops in an overmature tree that causes clusters of small white areas in the heartwood. Also called white pocket.

Laminated Strand Lumber (LSL)—Another name for engineered strand lumber.

Laminated Veneer Lumber (LVL)—A structural or nonstructural parallel assembly of veneer.

Lamstock—Special grades of wood used in constructing glulam.

Lapsiding—A board that has been resawn diagonally to be used to clad the exterior of a building. Also known as bevel siding.

Lignin—The second most abundant constituent of wood, after cellulose. It is the thin cementing layer between the wood cells.

Linga—(*Persea* spp) A tropical tree that grows from the West Indies and southern Mexico southward to Chile.

Machine Stress Rated (MSR) lumber—Lumber that has been mechanically evaluated to determine its stiffness and bending strength.

MBM—Thousand feet board measure.

Medium Density Fiberboard (MDF)—A dry-formed panel product manufactured from wood fibers combined with a synthetic resin or other suitable binder and compressed in a hot press to a density of 31–50 pounds per cubic foot.

Melamine—A formaldehyde-based adhesive often used in conjunction with other materials for decorative panel overlays.

MM—Million.

M^3—Million cubic.

Moisture content (MC)—The weight of the water in wood, expressed as the percentage of the weight of the oven-dry wood.

Nongovernmental organizations (NGOs)—Organizations that are privately sponsored, with no government affiliation.

Order file—A listing of the factory orders yet to be shipped to a customer.

Oriented Strand Board (OSB)—Panels constructed of thin wafers of fiber oriented lengthwise and crosswise in layers, with a heat-activated resin binder. Depending on the resin used, OSB can be suitable for interior or exterior applications.

Oriented Strand Lumber (OSL)—Lumber made of thin wafers of fiber in various configurations and bonded with a heat-activated resin binder.

OSB-based lapsiding—Lapsiding made from OSB.

Parallam—A proprietary name for a type of Parallel Strand Lumber.

Parallel Laminated Veneer (PLV)—A product in which the veneers have been laminated with their grains parallel to one another. A technique used in furniture and cabinetry to provide flexibility over curved surfaces. Laminated Veneer Lumber (LVL) is a close relative, although LVL is designed specifically for structural or select non-structural applications.

Parallel Strand Lumber (PSL)—A lumber product composed of parallel laminated strands of wood milled from dry veneer.

Particleboard (PB)—A generic term used to describe panel products made from discrete particles of wood or other ligno-cellulosic material rather then from fibers. The wood particles are mixed with resins and formed into a solid board under heat and pressure.

Permanent Wood Foundation (PWF)—A foundation system in which treated plywood and lumber products are used in place of concrete. PWF improves heating and cooling capabilities and can be installed in weather conditions that would prevent pouring of a concrete foundation. Originally called the All-Weather Wood Foundation, the PWF name was adopted in 1984.

Phenol-formaldehyde resin (PF resin)—Adhesive made from phenol compounds and formaldehyde used in the gluing of plywood and other composite panels.

Phenolic films—A thin PF-based covering applied to a manufactured panel for decorative purposes.

Playing to the statement—Making decisions that have a positive financial effect.

Preshredder—A machine that breaks up large pieces of wood into manageable sizes.

PrimeBoard—Proprietary name for particleboard that uses straw as the main ingredient.

Primfog—Proprietary name for a knot-free laminate.

Radiata pine—(*Pinus radiata*) A tree species that is widely planted in New Zealand, Australia, and Chile. In the United States it is commonly called Monterey pine.

Roundwood—Logs, bolts, and other tree sections produced from the forest.

R value—The resistance to heat flow of a material. The higher the R value, the more effective the insulation.

Sawlog—A log of sufficient size and quality to be suitable for lumber manufacturing.

Sawtimber—Timber suitable for processing into solid wood products such as lumber and plywood.

Scrimber—A whole-log technology formerly used by Scrimber International of Mount Gambier, South Australia. The log is debarked, the fiber is crushed lengthwise into bundles of interconnected and aligned strands prior to being manufactured into a lumber product.

Shiplap—Lumber that has been worked to make a lapped, or rabbeted, joint on each edge so that pieces may be fitted together snugly for increased strength and stability. Also, a similar pattern cut into plywood or other wood panels used as siding to ensure a tight joint.

Silvaculture—The theory and practice of forest establishment, composition, and growth.

SmartWood—A nonprofit company that certifies products and processes as ecologically sustainable, accredited by the Forest Stewardship Council.

Society of American Foresters (SAF)—Association whose purposes are to advance the science, technology, education, and practice of professional forestry and to use the knowledge and skills of the professional forester to benefit society.

Soffit—The underside of an eave or other part of a building.

Soil-site index—Soil properties used to predict tree growth. These would include the nutritional makeup of the soil and topogaphic and climatic features.

Springwood—The softer, more porous portion of an annual ring of wood that develops early in the growing season.

Structural Composite Lumber (SCL)—A wood composite substitute for solid sawn lumber, timbers, and beams.

Substrate—A layer of material that supports something applied to its face.

Summerwood—The dense fibrous outer portion of each annual ring of a tree, formed late in the growing season, although not necessarily in the summer. The opposite of springwood.

Surface checking—Checking that occurs on the surface.

Sustainable Forestry Initiative (SFI)—A second-party ecological verification system developed and used by the members of the American Forest & Paper Association (AF&PA) and their contract suppliers.

Thermomechanical newsprint—Newsprint made by a process in which the wood chips are heated and softened by steam before being ground into pulp.

Thermoplastics—Material that is capable of being repeatedly softened by heat and hardened by cooling. Some glues and resins used in the wood industry are thermoplastic, although most are thermosetting.

Thermosets—Material, such as an adhesive, that sets when heat is applied. A thermosetting glue does not soften when later subjected to heat.

Timberstrand LSL—Proprietary name for OSL/LSL in North America.

Trex lumber—A proprietary lumber product that utilizes wood fiber and consumer plastic waste as its primary raw materials.

Triboard—Proprietary wood composite panel that integrates Medium Density Fiberboard (MDF) and OSB technology. The panel is constructed from a core of strands sandwiched between surface layers of fiber. Steam injection pressing allows the construction of a thickness in excess of 4 inches (about 100 mm).

Urea formaldehyde resin (UF resin)—Adhesive with a high molecular weight made from amino acids and formaldehyde used in the gluing of plywood and other composite panels. It contains more formaldehyde than phenol-formaldehyde adhesives.

Urban forest—1. The pallets, wood debris, and other like materials recovered and recycled in major population centers. Also, aged wooden structures whose components are recycled, such as old-growth beams, timbers, and other lumber products. Also called ancient timber. 2. Trees grown in an urban setting.

Utility white spec veneer—A grade of fir veneer that allows white speck and more defects than are allowed in D grade.

Vertical grain—Wood grain sawn at right angles to the annual growth rings so that the rings form an angle of 45 degrees or more with the surface of the piece.

V grooving—Marking of a panel or lumber product in a "V" shape to enhance its appearance.

Waferboard—A panel product, usually considered the forerunner of OSB, that does not have wafer layers that are perpendicular to each other. Waferboard has lower strength values than OSB.

WWF 1995 Plus Group—A large, developed buying group located in the United Kingdom.

Wood Residue Receiving Centers (WRRC)—Centers set up to receive wood-fiber-based recyclable materials.

Index